Sustainability Activities We All Can Do To Help Save The Planet

R.L. Gemeinhardt

Other Books By R.L. Gemeinhardt available from Amazon

- Earth Dwellers Guide To Recycling And Environmental Conservation For Kids And Teachers

- Earth Dwellers Guide To Recycling And Environmental Conservation For Companies

- Earth Dwellers Guide To Recycling And Environmental Conservation

- DEMOB THIS!!! A Jack Owens novel about his response to the oil spill from hell!

- Texas Used Oil Management: A Practical Guide To Compliance

- Author web site http://rlgemeinhardt.com/

COPYRIGHT & CREDITS

Text copyright © 2018 R.L. Gemeinhardt.
Cover art copyright © 2018 R.L. Gemeinhardt.
All rights reserved.

No part of this publication may be reproduced, distributed, or transmitted in any form or by any means, including photocopying, recording, or other electronic or mechanical methods, without the written permission of the copyright owner, except in the case of brief quotations embodied in reviews and certain other non-commercial uses permitted by copyright law.

Requests for authorization should be addressed to VRM Group, LLC
Attn: Ron Gemeinhardt
190 — B2 Gulf Freeway, #126
League City, TX 77573

ISBN- 9798667439271

Limit of Liability/Disclaimer of Warranty: While the publisher and author have used their best efforts in the publication of this work, neither author nor publisher makes any representations or warranties with respect to the accuracy or completeness of the contents of this work and specifically dis- claim any implied warranties of merchantability of fitness for a particular purpose. No warranty may be created or extended by sales representatives or written sales materials. The advice and strategies contained herein may not be suitable for your situation. You should consult with a professional where appropriate. Neither the publisher, nor the author shall be liable for any loss of profit or any other commercial or personal damages, including but not limited to special, incidental, consequential, or other damages.

DEDICATION

To all those people in my life both personal friends and business acquaintances that have shown me so many times that any positive environmental change depends on actions we take as individuals is the only real way to impact meaningful change to this planet we all share.

"Never doubt that a small group of thoughtful, committed citizens can change the world; indeed, it's the only thing that ever has."

— MARGARET MEAD

PREFACE

"The Earth does not belong to us: we belong to the Earth."
~ MARLEE MATLIN

This book contains sustainability activities you can do as an individual or a family to help save the planet. It is really not hard for each of us to do just a few things in our personal lives that would really help in this battle we are all in to insure our planet continues to support us as humans. This book tries to serve as a call to action to get you and your kids involved and can even help inspire family and friends to find their own path to sustainable living that reduces the damage of modern life on the planet.

First a couple definitions you are probably familiar with but I would like to revisit so you know my usage as well.

Sustainability: It is a word many people have heard but maybe don't really understand. Frankly, many definitions exist out there all with the same focus but the one I like from an overall understanding and application to our lives is from the U. S. Environmental Protection Agency (USEPA):

"Sustainability is based on a simple principle: Everything that we need for our survival and well-being depends, either directly or indirectly, on our natural environment (*definition below*). To pursue sustainability is to create and maintain the conditions under which humans and nature can exist in productive harmony to support present and future generations."

Natural Environment: All living and non-living things occurring NATURALLY, meaning in this case not artificial. This ENVIRONMENT surrounds and has interaction with all living things as climate, weather and NATURAL resources; soil/rock, air, water, sun all part of the physical environment that affect human survival and economic activity.

This book tries to serve as a call to action to get you and your kids involved and can even help inspire family and friends to find their own path to sustainable living that reduces the damage of modern life on the planet. If you need a why, just some of the impacts of our current life styles are listed below:

- Every year, we extract an estimated 55 billion tons of fossil energy, minerals, metals and bio mass from the earth.
- The world has already lost 80% of its forests, and we continually lose more at a rate of 375 km2 per day!
- At the current rate of deforestation, 5-10% of tropical forest species will become extinct every decade.
- Every hour, 1,692 acres of productive dry land become desert.
- 27% of our coral reefs have been destroyed. If the existing rate continues, the remaining 60% will be gone in 30 years.

The numbers may look daunting, but many seemingly minor changes in lifestyle, when adopted on a wide scale, can make a significant difference. From simple solutions for paper waste to plastic bags and bottles all the way to rethinking transportation and household appliances, and even making a delicious vegetable stock, we all have countless opportunities to conserve, recycle, and expand our positive impact exponentially by sharing methods that work.

In each section, you'll find not only useful information and ironclad reasoning for positive change, but also detailed advice to turn knowledge into action. There are already plenty of resources that take a scholarly approach; instead, I have tried to present user-friendly applied methods that suit your daily life. So if you want to know what you can do to save the world, in baby steps or big strides, read on for ways to change your life and, through that, the planet's. Even if you're not yet ready for a deep dive into sustainability, recycling and conservation just doing a few of the options provided is an excellent start.

"The greatest threat to our planet is the belief that someone else will save it."
— ROBERT SWAN, AUTHOR

The first step is getting informed. Then, whatever you do, **JUST DO SOMETHING!!!**

Table of Contents

	Page
Preface	5
SECTION ONE: STARTING OFF	10
1 - Thermostat Control	11
2 - Water Use	15
3 - The Problem of Plastic Bags	24
4 - Aluminum Cans	31
5 - Disposable Bottles	35
6 - Paper	41
7 - Styrofoam	47
8 - Litter	55
9 - Wood and Wood Furniture	60
10 - Buy a Rain Barrel	63
SECTION TWO: MOVING FORWARD	71
11 - Batteries & Solar Chargers	73
12 - Light Bulbs	84
13 - Appliances	94
14 – Options For Less Driving	97
15 - Used Motor Oil	102
16 - Plant a Tree	110
SECTION THREE: ADVANCED ACTION	116
17 - Buy Local	117
18 - Sustainable Wood	125
Conclusions	139

Table of Contents

	Page
References	145
Notes / Web Site References	157
About The Author	163

SECTION ONE: STARTING OFF

"The only way forward, if we are going to improve the quality of the environment, is to get everybody involved."
~ RICHARD ROGERS

Over my years of educating people and organizations about regulatory compliance, recycling, conservation and sustainable practices, I've heard many reasons why people put off making changes in their routine. The biggest arguments are that being environmentally friendly costs too much time and money. Fortunately, the reality is the opposite. Sure, change takes a little effort, at first. But with a few simple actions, we can break generations of bad habits and form better ones to teach our kids. What we dismiss in the moment out of minor inconvenience makes a major difference over the life of our planet.

And, here's the best part: you can save the planet and your money.

In this first section, I'll focus on small changes that help build good habits while reaping huge rewards for the environment and your wallet. Not only can sustainability support activities save you money, but also make you some in the short and long term. So if you're worried a shift to sustainable living will be overwhelming, try a few of these small steps to get you started and build our healthier environment as we help save the planet one positive change at a time.

1
Thermostat Control

Sustainability often involves the reduction in your consumption of natural resources such as fuel or energy in various forms in our day to day lives. As this first activity may present a little comfort challenge it can actually pay big dividends as we try to more closely regulate our homes' thermostats. But the good news is: with only a little adjustment, a few minor changes in routine can do a lot of good and save you money, too.

WHAT YOU CAN DO:

Some facts about saving money from the Department of Energy (DOE):

DOE estimates savings[1] of about 1% for each degree of thermostat adjustment per 8 hours, and recommends turning thermostats back[2] 7 to 10 degrees from their normal settings for 8 hours per day to achieve annual savings of up to 10%.

- You can save as much as 10% a year on heating and cooling by simply turning your thermostat back 7°-10°F for 8 hours a day from its normal setting. The percentage of savings from setback is greater for buildings in milder climates than for those in more severe climates.

From a personal perspective, my family's monthly electrical bill ran approximately $250 per month in a 2800 sq. ft. house in Texas. Following this process, we actually saved about 8% or about $20 a month, close to $240 a year — like getting a month free. It takes some getting used to, but with a programmable system, we found it pretty easy to set up.

Just imagine: if ten percent of the population made this one simple change, the saving in energy use would be more than a little significant.

Save energy and money in four simple steps with a programmable thermostat. The simplest method to achieve maximum energy savings with temperature control is by installing a programmable thermostat. Once it's set, your job is done. Here's how a weekday schedule might look for a family with adults and children that are out of the house all day for work and school:

1. 6:45 a.m.: The family wakes up to get ready for the day. The temperature of the house is 68°F; the heat automatically turned on a bit earlier so it would hit this temperature by 6:30.
2. 7:45 a.m.: The family leaves the house and the thermostat is set to 56°F. By turning their thermostat back 10° to 15° for 8 hours, the family can save 5% to 15% a year on their heating bill—a savings of as much as 1% for each degree if the setback period is eight hours long.
3. 4:30 p.m.: The family starts returning home from work and school. The heat turned back on a bit beforehand so the house would again be 68°F for their return.
4. 10:30 p.m.: The whole family has gone to bed (bundled in warm pajamas and snuggled under blankets), and the thermostat is again set to 56°F.

What if I don't have a programmable thermostat?

If you rent or live in an apartment, you might not be able to install a programmable thermostat. Or, maybe you're not ready to make such a big change yet. Just make adjustments manually and set some daily reminders to get into the habit of turning your thermostat down during the day and overnight. With only a little extra effort, you can still see plenty of impact.

Set your summer energy savings by degree:

It's estimated that for every degree higher you set your thermostat over 78 degrees Fahrenheit in the summer months, you could save approximately 6 to 8% off your home energy bill. Yes, you read that correctly: six to eight percent savings per degree.
Try the following settings:
1. 78° F when you're home.
2. 85° F when you're at work or away.
3. 82° F when you're sleeping.

If you're more heat-tolerant, you can experiment with the temperature, raising it 1 degree at a time to see how it affects your comfort and your budget; the savings still adds up pretty quickly. And, if you aren't comfortable at 78° F, lower the temperature a degree at a time and let your system reach the new setting before ratcheting it down further.

To make the most of your savings:

- Clean or replace air filters in your air conditioning unit at least once a month.
- If you have central air conditioning, close vents in unused rooms.

But I have a window air conditioner:

If you don't have central air and depend on window air conditioners, it's more difficult to reach the perfect temperature. Because the thermostat is in the unit itself, it measures only the temperature in that part of the room. So, depending on the size and shape of the room, the thermostat may not provide a consistent temperature throughout the space you want to cool.

That means getting the right comfort level requires a little trial and error.

- Start with it set at 78° F and see how you feel.
- If you have a window unit in your bedroom, wait until 30 minutes or so before going to bed to turn it on so you're not spending energy cooling an empty room.

Share the wealth:

It may take you some convincing to bring your entire household on board, but share with them the cost savings actually achieved. Give everyone a stake in your success, make a plan together, and promise a reward when your team effort pays off. Then, I hope you'll also share the advantages of reducing your personal carbon footprint with extended family and friends to expand the benefits to our environment.

2
Water Use

Water use is one of the significant sustainability issues as we really only have a limited supply in all areas where we live and minimizing individual use as well as overall family use while still insuring a healthy lifestyle can be done with minimal effort and again potentially save you money. Yes, water use is another activity with some impact on creature comfort. I'll identify simple controls and methods of water and energy savings, from tips you can immediately implement in your daily life to selecting fixtures and appliances that increase your savings even more. By just picking a few of the suggested changes—maybe even all—you can really make a big difference.

WHAT YOU CAN DO:

General consumption tips:

Here are a few general conservation tips with some background information from the DOE's National Renewable Energy Laboratory. Whether you're looking for no-cost habit changes, low-cost purchases or improvements, or large investments like new water heaters or appliances, there's a useful tip in here for you.

General use:

Fix leaks:
A leak of one drip per second can cost $1 per month. That twelve-dollar-a-year delay in repairs may not seem like a big deal, but the American Water Works Association has put the extended impact into perspective.

If you hold off fixing that "little" leak, at 60 drips per minute, you'd waste:

- 8.64 gallons per day
- 259 gallons per month
- Just over 3,153 gallons per year

That's a LOT of good, clean water plus a piece of your budget lost.

Don't let the water run:
Do you leave the water on while brushing your teeth? Turn on the shower and let it warm up while you get towels or a washcloth? Or maybe you step away from the kitchen sink to grab dirty dishes or get a scouring pad? These seemingly minor actions can greatly impact overall water use. All of those little moments can add up to a lot of wasted water. Try shutting the water off while not in active use, and see how great an impact your mindfulness makes to your water and water-heating bill.

Install low-flow fixtures:
What if I told you that, for a small investment and a few moments of effort, you can achieve water savings of 25%-60%? Showerheads and faucets that pre-date 1992 can use more than twice as much water as new ones. That's why federal regulations require new showerheads and faucets to have low flow rates. Once installed, low-flow fixtures will do the work in reducing your consumption and water bills. Your local hardware store should be able to help you identify specific fixtures for your needs.

Showers & bathing:

Take short showers instead of baths:

Your capacity for savings here will depend on you and your household's habits. A long, hot shower may actually use more hot water than a bath not filled to the brim. But even a bath filled to only a few inches can use a heck of a lot of water if you have a large tub. Soaking in the bath is a nice luxury, now and then, but for daily bathing stick with a short shower for the best water efficiency. And, for extra water savings, try turning off the water while soaping up, shampooing, or shaving.

Reduce your time in the shower:
I know; a nice long shower feels great, especially on a chilly morning. Even I sometimes spend too long in the shower because I'm just so comfortable and it's too cold to want to get out. This is anecdotal, but keeping the bathroom door tightly closed when showering seems to keep the air much warmer; just run the fan to take care of the steam. Keep a big towel and warm fluffy robe nearby to kill your chill.

Laundry & dishwashers:

Use cold water for most laundry:
A lot of people believe all soaps need hot water, which is simply not true. Most laundry loads will clean just fine in cold water. And, for the few exceptions where a hot wash is most advisable—like kitchen rags and towels or baby's whites—you can still always use cold water for the rinse cycle. You'll likely notice no difference in the end result except for some tidy savings in your water-heating bill.

Use your dishwasher efficiently:
The dishwasher is a great modern convenience, but a real drain on conservation efforts when used inefficiently. To maximize your water usage:

- Wash only full loads.

- Choose shorter wash cycles.
- Activate the booster heater if your dishwasher has one.

Install an ENERGY STAR®-qualified dishwasher:
Here's another no-brainer for folks who want to maximize their conservation efforts and savings. The best part is, once your new dishwasher is installed, most of your work is done. Each load, without lifting a finger, you'll use 31% less energy and 33% less water. That's a pretty good reason to consider making the investment in switching. Then, plan for full loads only to enjoy even more savings.

Upgrade your clothes washer:
ENERGY STAR® says that you could fill three backyard swimming pools with the water you save over the life of a new ENERGY STAR®- qualified washer(3). That staggering figure, alone, is plenty of reason to consider an upgrade. But here's another you might like even better: if you're replacing a washer that's over 10 years old, you can save over $135 per year.

Water heaters:

Help your water heater save you money:
For maximum savings, you can increase the energy efficiency of your water heater and hot-water storage tank. Just try these few tips:

- Insulate your hot-water storage tank.
- For electric tanks, be careful not to cover the thermostat.
- For natural gas or oil hot water storage tanks, be careful not to cover the water heater's top, bottom, thermostat, or burner compartment.
- Always follow the manufacturer's recommendations.
- Insulate the first few feet of hot and cold pipes

connected to the water heater.
- Install a timer to turn off your electric water heater at night or when it's not in use.
- Use the timer to turn off the water heater during your utility's peak demand times.

Purchase a new water heater:

Sometimes, we can get stuck in a rut of viewing unseen workhorses like our water heaters through the lens of what feels familiar. The surge in environmentally friendly innovation has made tremendous advances to old-school fixtures that hadn't changed much over generations. Don't limit yourself to just conventional storage water heaters. Other more efficient choices are available that might be right for you, once you learn more about your options and considerations.

Check your fixtures and appliances:

Below are some specifics on how to check out various appliances and devices to consider what's best for you and your family, using summarized DOE information with some additional comments based on my experience.

Showerheads:

Are you using low-flow?

For maximum water efficiency, select a showerhead with a flow rate of less than 2.5 gallons per minute (GPM). Before 1992, some showerheads had flow rates of 5.5 GPM. Therefore, if you have fixtures that pre-date 1992, you might want to consider replacing them if you're not sure of their flow rates.

Here's a quick test to determine whether you should replace a showerhead:
1. Place a bucket — marked in gallon increments — under your showerhead.

2. Turn on the shower at the normal water pressure you use.
3. Time how many seconds it takes to fill the bucket to the 1-gallon (3.8 liter) mark.
4. If it takes less than 20 seconds to reach the 1-gallon mark, you could benefit from a low-flow showerhead.

I tried this test recently and was surprised, as I thought we had up-to-date fixtures. The initial shock, fortunately, was offset by the unexpected chance to put in some fancier fixtures we'd been considering.

Which low-flow showerhead is right for you?

There are two basic types of low-flow showerheads:

- Aerating showerheads mix air with water, forming a misty spray.
- Laminar-flow showerheads form individual streams of water. If you live in a humid climate, you might want to use a laminar-flow showerhead because it won't create as much steam and moisture as an aerating one.

Dishwashers:

Do you wash by hand?
It's commonly assumed that washing dishes by hand saves hot water. However, hand-washing several times a day can actually be more expensive, especially compared to an energy-efficient dishwasher. When combined, the water, energy and time savings make a sound argument for investing in a new dishwasher. You can consume less energy with an energy-efficient dishwasher when properly used with only full loads. Of course, full loads can be trickier for smaller households, who use fewer dishes overall, but rarely a problem for most families.

Thinking of getting a new dishwasher?

Consider Capacity:

When purchasing a new dishwasher, always check the Energy Guide label to see how much energy it uses, but also consider capacity. Dishwashers fall into one of two categories:

- Compact-capacity dishwashers can appear to be more energy efficient on the Energy Guide Label, but they hold fewer dishes, which may force larger households to run it more frequently.
- Standard-capacity dishwashers might seem to have higher energy costs, but in many cases can actually save energy and costs in the long run. While some might not need the room, in larger households the space for a few extra dishes can prevent the waste of running additional loads to finish regular cleanups.
- **Use a booster heater:**
  The booster heater makes a dishwasher more energy efficient by increasing the temperature of the water entering the dishwasher to the 140°F recommended for cleaning. Since your best energy savings happen when your water heater temperature is set to 120°F, you'll need this increase in water temperature to get everything clean. Some dishwashers have built-in heat boosters, while others require manual selection before the wash cycle begins or only activate the booster during the heavy-duty cycle. Dishwashers with booster heaters typically cost more, but pay for themselves with energy savings in about 1 year if you also lower the temperature on your water heater.

Cycle selections:

Another important dishwasher feature is cycle selections. This allows you to adjust the length of a washing cycle based on the needs of each load. Shorter cycles require less hot water, thereby reducing both water waste and energy cost.

Choose ENERGY STAR®

This choice is really a no-brainer. If you want to ensure that your new dishwasher is energy efficient, purchase one with an ENERGY STAR® label. Once it's installed, you'll enjoy savings of approximately 33% in water usage and 31% in energy consumption on every load. Always run only full loads to maximize your impact.

Clothes Washers:

Here are a few tips to start saving more money and do the planet some good by bringing more efficiency to your heated water management and minimizing water use.

Temperature:

Unlike dishwashers, clothes washers don't require a minimum temperature for optimum cleaning. Therefore, to reduce energy costs, you can use cold water for most laundry loads. Cold water is always sufficient for rinsing.

Energy efficiency:

Inefficient clothes washers can cost three times as much to operate than energy-efficient ones. Of course, I wouldn't expect you to go out and get a new one to replace a perfectly competent washer. But if you've got an older model, at least use the cold water cycle.

If you need a new clothes washer, consider getting a more energy-efficient one. Another feature to look for in a new machine is the ability to adjust your water temperature and levels for different loads. Efficient clothes washers spin-dry your clothes more effectively, too, saving energy when drying as well. Plus, look into a front-loading machine, which uses less water and, consequently, less energy than top loaders.

Capacity:

Small-capacity clothes washers often have better Energy Guide label ratings. However, a reduced capacity might increase the number of loads you need to run, which could increase your energy costs. When shopping for a new machine, make sure to weigh your actual usage against the benefits of reduced capacity; you might find that a larger-capacity machine would actually be better for your overall needs.

Toilet:

The first, and most obvious solution to water waste is replacing an older toilet with a newer water efficient one. But, even if you can't replace your older model, an easy way to reduce water consumption per flush is to displace some of the water in the tank with a brick. This simple trick allows you to get the same flush pressure, but use up to a half-gallon less water per flush.

Okay, maybe you're worried about crumbling brick in your toilet; instead, you can put a plastic water bottle in toilet tank. With a plastic bottle — or even bottles — that you've stopped using (since you carry water in a reusable bottle instead of purchasing disposables, right?), you can achieve the same results without worrying about brick pieces clogging your toilet.

1. Put a few pebbles or rocks in the bottle.
2. Fill it with water.
3. Place it in the back of your toilet.
4. You just saved up to 10 gallons of water per day.

3
The Problem of Plastic Bags

If you really want to make only one specific change out of all the sustainability activities offered here this would be the one to do because eliminating or at least reducing the plastic bags in your life can have a tremendous impact on our environment. Not trying to be over dramatic, but it would really be a great victory for the earth and all who live on it. I know it's difficult to imagine life without plastic bags. They've become so much a part of our routine that the average American throws away about ten single-use plastic bags per week. This may not seem like a lot, but it adds up fast. Americans use 100 billion plastic bags a year (4) that wind up in landfills or as litter, leeching toxic chemicals, killing critical wildlife, and damaging ecosystems. And the plastic problem is global.

In this activity I'll first show you some ways you can reduce your plastic bag usage. Then, I'll share some disturbing facts about the size and extent of the problem and its impact. Once you've tackled your plastics consumption, I hope you'll use the information to inspire family and friends to do the same. Together, we can make a real difference.

WHAT YOU CAN DO:
Eliminating plastic bag waste isn't a complicated matter. It requires no step-by-steps or number crunching. All you need is the resolve to change a few habits and mindfulness to look for more earth-friendly solutions.

Here are a few places to start:

- To begin your conversion to a non-plastic bag household, you'll need cloth/cotton shopping bags.

-
  - a) Cloth/cotton bags are available at most general retailers like Target, Costco, and Walmart as well as some supermarkets and online suppliers.
  - b) Look for roomy bags with long, comfortable handles for easier carrying
  - c) Get at least a dozen to have some at home and back-ups in your car(s) for last-minute shopping trips.
- Don't forget cotton string bags for your produce.
  - These are also readily available at many retailers and online shops.
  - To determine the right number for you, make a list of what produce items you buy on an average shopping trip and add a few more for larger grocery runs.
- Buy in bulk for less packaging waste per serving.
  - a) For pre-packaged goods, avoid single-serve bundles.
  - b) Many dry goods are available in bulk at greatly reduced prices.
  - c) When possible, bring your own container or reuse your own plastic bags for regular bulk dry goods purchases.
- Whenever possible, just say no.
  - a) Larger or singular produce items can be packed without the extra bag.
  - b) When shopping for smaller or single items at any store, ask to carry your purchase without the bag.
- Don't limit your cloth/cotton bags to just groceries. Bring them along for all your shopping trips to maximize your impact. Many smaller purchases of non-grocery items don't require bagging. If you must take a store-supplied bag:
  - a) Take the paper option whenever possible. Even if you eventually dispose of it, instead of sitting in that landfill for five hundred years, it'll

biodegrade in just
- b) Take the paper option whenever possible. Even if you eventually dispose of it, instead of sitting in that landfill for five hundred years, it'll biodegrade in just a few weeks.
- c) Reuse bags as often as possible before disposing.
- d) Recycle used bags if you can.

WHY IT MATTERS:

Some background:

Plastic bag abundance has not always been the problem it is today; until the late 1970s, single-use plastic bags were seldom available in grocery stores. Since then, an estimated one trillion bags are used each year globally. And the biggest problem is: they're so seamlessly ingrained into our daily routines, we hardly notice. While their environmental costs are burden- some for communities and the planet, the financial cost of plastic bags for retailers is pretty low. Made from ethylene, a byproduct of petroleum or natural gas, plastic bags are so cheap and flimsy that cashiers use them freely, double bagging as a matter of course and often sticking only a few items in each bag. As a result, shoppers end up with piles of plastic bags.

New Yorkers alone use twice the national average. Some twenty-three billion are used across the state [5] each year — more than enough, when tied together, to stretch to the moon and back thirteen times. The United Nations Environment Program estimates[6] that eight million tons of plastic waste that includes plastic bags end up in the oceans each year, while a 2016 report by the World Economic Forum report projects[7] there will be more plastic than fish by weight in the oceans by 2050 if current trends continue. At the same time, the plastic production and disposal also generates around four hundred million tons of carbon dioxide a year globally [8].

If you fish, hunt or just enjoy the outdoors you should know that millions of whales, birds, seals and turtles have been killed because they mistake plastic bags for food or become ensnared in packing bands and other items. Trillions of micro plastics (9) end up in the ocean, with seafood eaters ingesting an estimated 11,000 tiny pieces (10) annually. If you think it is just nature's problem, you should know that plastic fibers have also been found (11) in tap water around the world; in one study, researchers found them in 94% of water samples in the United States.(12)

Facts and harmful impacts of plastic bags:

Here are some raw facts about plastic bags and their harm to our planet:

General impact and consumption:

- It can take five hundred (or more) years (13) for a plastic bag to degrade in a landfill. While this estimate may be a worst case scenario, as many bags can take five to ten years to decompose, either way it's a long time.
- Decomposition doesn't mean the chemicals go away, just that they're not in bag form anymore. Unfortunately, the bags don't break down completely but photo-degrade, becoming micro plastics that absorb toxins and continue to pollute the environment.
- Americans use 100 billion plastic bags a year (14), which require twelve million barrels of oil to manufacture. So, beyond the pollution and harm to humans, wildlife and our waters using plastic bags requires a lot of oil.
- The fossil-fuel equivalent to produce only fourteen plastic bags (15) would provide the gas to drive one mile.

- Plastic bags are used for an average of twelve minutes (16). Think about all the bags that gather in your pantry if, like me, that's where you store the bags you can't avoid taking. Even though we try to reuse them, at some point they're put into the recycle bin. We're lucky that in our area they do actually take them, but they still stay buried someplace.
- The average American family takes home almost 1,500 plastic shopping bags per year (17). I encourage you to do the math for your family, and don't forget to count all sources, not just your regular grocery store trips.
- According to Waste Management, only one percent of plastic bags are returned for recycling (18). That means the average family only recycles fifteen bags a year; the rest end up in landfills or as litter. When you also add that some municipalities have stopped allowing them in their recycling programs, the issue becomes more catastrophic. What do they expect us to do with all those bags, eat them? With the exception of California and a few others that have banned single-use plastic bags in many areas, most governments and their agencies have failed in the critical effort to reduce plastic waste.
- Plastic bags are a source of litter on land and in waterways and avoidable packaging waste in landfills that are only used for mere minutes.
- They create tangles and jams in recycling and waste water processing equipment.
- They're costly for municipalities and recycling centers to recycle, in terms of time and money, and in many cases municipalities have stopped taking plastic at all.

Harm to oceans and wildlife:

- Up to 80% of ocean plastic pollution [19] enters the ocean from land.
- At least 267 different species [20] have been affected by plastic pollution in the ocean.
- Estimates project that 100,000 marine animals [21] are killed by plastic bags annually.
    a) Fieldwork from Wallace "J." Nichols, a marine biologist and research associate for the California Academy of Sciences, identifies that while a lot of figures have been thrown around in the media, hard numbers are very difficult to calculate. The sad fact is: when most sea animals eat plastic and die, they sink to the bottom, unaccounted for.
    b) Possibly even more significant are animals affected indirectly. For example, when sea turtles eat plastic instead of food, their glucose levels drop, leaving them with less energy for migration and reproduction. Females can't lay as many eggs, and fewer new sea turtles are born. "When you connect the dots," Nichols said, "you realize that plastic pollution may cost millions of potential sea turtle lives."
    c) One in three leatherback sea turtles [22] have been found with plastic in their stomachs.
- It is not just an American issue. A Google search for "animals eat plastic bags" brings up hundreds of heartbreaking stories and images from all over the world. Large numbers of foraging cows in India have died from ingesting plastic bag litter, even though in that country they've banned the distribution of plastic bags. In the United Arab Emirates, a veterinarian has documented images of camels, sheep, goats, and endangered desert animals that die from eating plastic

bags. Whales wash up on coasts throughout the globe, their bellies full of plastic.

Plastic bags and human health:

- Plastic bags aren't just a problem impacting water and aquatic life; plastic affects human health.
    a) Toxic chemicals leach out of plastic and are found in the blood and tissue of nearly all of us.
    b) Exposure to them is linked to cancers, birth defects, impaired immunity, endocrine disruption and other ailments.
- Below are two videos I strongly recommend, one for you and one for the kids.

    a) "Peril of Plastics: risks to human health and the environment" (2010) from a professor at Arizona State University identifies issues of concern and is well worth your time. It can be found at this link: https://biodesign.asu.edu/news/perils-plastics-risks-human-health- and-environment
    b) The second recommendation is a PBS video available on YouTube, part of the Nature Cat Series that addresses plastic bags at a kids' entertaining level. I suggest you watch it with your kids; it would be great time to talk with them about assisting in the effort to curb your plastic bag use. The video can be found at: https://www.youtube.com/watch?v=Wi99ypudQ3c

4
Aluminum Cans

To start this activity off, I want to suggest you go to a video from Chula Vista Clean & Green Curiosity Quest (found at: https://www.youtube.com/watch?v=jHx95PQIl4k) you can view with your kids to show how easy it is to recycle your aluminum cans. It's a great kid activity that can be a lot of fun and lets them get more involved in your conservation efforts. Plus, you can actually make a few bucks for doing the right thing. And, if you want to really motivate your younger generation and help teach a great lesson in responsibility, let them manage the recycling all by themselves and earn extra money to spend any way they like.

What you can do:

1. Collect the cans in a separate clean container or plastic bag.
   a) Make sure cans are empty prior to storage to avoid attracting bugs while they're stored.
   b) You don't have to rinse, but the tiny extra step when you finish a drink will ensure no lingering sweet residue.
2. I suggest you periodically squash the cans, to fit more in your container and take less frequent trips to the recycler.
   a) Can flattening is a great stress relief and even fun for the kids, but be safe.
   b) There are also manual and electric helpers that work well, too.
3. You can find a recycler in your area on the web, or even call your state agency and they can tell you the location of your closest recycler.

4. Gather up the cans, collect the kids, and take them with you to show them the end of the process.
5. Turn in your cans and collect your money!

How much money can I really make?

Here are some figures to help you determine the potential earnings for recycling.

Currently, in the U.S., eleven states and Guam have active bottle bills, which require minimum payouts of between 5 cents and 10 cents per recycled aluminum can. This means that if you live in one of these areas, your income potential for recycling aluminum cans just went up astronomically. They are:

- California
- Connecticut
- Delaware
- Guam
- Hawaii
- Iowa
- Maine
- Massachusetts
- Michigan
- New York
- Oregon
- Vermont

In states with no bottle bill, prices go up and down but are usually similar throughout the country at any given time. With approximately a half-ounce of aluminum per can—or 29 cans per pound—and a $0.60 per pound market rate for general aluminum, the least you should receive for your cans is $0.30 per pound. Anything less than that is likely a scrap yard dealer attempting to maximize their profits at your expense.

Think about it; it won't take long to collect a few hundred cans.

- Average households generate about 350 cans per person, per year.
- A family of 4 can generate at least 1400 cans per year, not counting parties or home gatherings. So, conservatively, a family could generate 1800 cans per year.
- At a going rate of $0.60 per pound and 1800/29 gives you approximately 62 pounds times $0.60 = $37.00 at a minimum.
- But, if you live in one of the states that have bottle bills, you could make $90 to $180 a year, which is not bad at all.

Take that money and spend it wisely or give it to the kids to buy something they really want as a nice reward for all their effort to help save the planet.

Why it matters:

Aluminum recycling benefits the environment and future generations in numerous ways. Here are a few facts:

- Aluminum recycling saves an estimated 95% of the energy required for aluminum production from ore, greatly reducing air emissions including greenhouse gases.
- Alumina, the raw material for primary aluminum production, is extracted from bauxite. For every pound of aluminum recycled, four pounds of bauxite are conserved.

- Over 50 million pounds of aluminum cans are recycled every week, saving critical space in landfills. An aluminum can that's thrown away will still be an aluminum can 500 years from now.
- Because so many aluminum cans are recycled, they account for less than 1% of the total U.S. waste stream, according to EPA estimates.
- YET, Americans throw away enough aluminum every month to rebuild our entire commercial air fleet. On average, Americans drink one beverage from an aluminum can every day. But we recycle just over 49% of the cans we use.
- A steel mill using recycled scrap reduces water pollution, air pollution, and mining waste by about 70 percent.

5
Disposable Bottles

Americans' disposable bottles, alone, use approximately one barrel (42 gal) of crude oil per second, which adds up to 86,400 barrels a day. This doesn't count the natural gas and fuel involved in the manufacturing process and transportation to get them to you.

Plus, disposable plastic bottles contain chemicals that can cause a litany of hazards to human, animal, and environmental health. These include: cancer, neurological difficulties, reduced fertility, reduced sperm count, testicular abnormality and tumors.

Obviously, this isn't a sustainable habit for our bodies, resources or our planet.

WHAT YOU CAN DO:

- Carry a BPA-free reusable bottle instead of purchasing drinks on the go.
- Avoid buying drinks in plastic bottles when possible, and recycle them.
- Opt for recyclable paper cartons or cups when you can.

Really. That's it. Much of the harm wrought by disposable plastic bottles could be mitigated by just a little forethought and willingness to accept a minor inconvenience.

WHY IT MATTERS:

By the numbers:

Increased use of plastic bottles has driven a critical issue to catastrophic levels. Between energy, resource and water consumption, we pay a dear price for the convenience of on-the-go drinks. But our kids will pay a far greater one if we don't act now to change our habits.

- In the U.S., 1,500 plastic water bottles are consumed every second [23].
- Manufacturing a quart (liter) plastic bottle requires 2.6 oz. (100ml) of crude oil.
- If you fill a plastic bottle with liquid so that it's 25% full, that's roughly how much oil it took to make the bottle.
- At the current estimated rate of 1500 bottles per second, Americans use approximately 86,400 barrels of oil a day for just disposable plastic bottles.
- Making a quart bottle also requires 1.5 cubic feet (42 liters) of natural gas. That's a sphere 53 inches (135 cm) wide.
- China, USA, Mexico, and Indonesia [24] are the four largest consumers of bottled water.

The Human Impact:

Bisphenol A (BPA):

Here are some facts from "One Green Planet" on plastic bottles containing Bisphenol A (BPA) [25] that aren't necessarily intended to scare you. But, I hope they help inform you about stuff you should be aware of and make you think about switching to a hard plastic bottle (sports bottle) with filtered water from your own home instead of supposedly recyclable plastic bottles. One qualifier is that the information applies to most "plastic" disposable bottles, and you should actually verify what you're using, because of the possible exposure issue identified below.

Plastic bottles contain Bisphenol A (BPA) (26), the chemical used to make the plastic hard and clear. BPA is an endocrine disruptor, which has been proven to be hazardous to human health (27). It has been strongly linked to a host of health problems including:

- Certain types of cancer
- Neurological difficulties
- Early puberty in girls
- Reduced fertility in women
- Premature labor
- Defects in newborn babies

Although several scientific studies have been done concerning the problems of chemicals found in bottled drinks, there have been various campaigns to undermine the results of the research. The American Chemical Council (ACC) still claims that BPA is safe.

BPA enters the human body through exposure to plastics, such as bottled drinks and cleaning products. It has been found in significant amounts in at-risk groups such as pregnant women's placentas and growing fetuses (28). A study conducted last year found that 96% of women in the U.S. have BPA in their bodies.

The good news is that you can have your BPA levels measured and make lifestyle changes to lower them, as demonstrated by Jeb Berrier in his film about plastic consumer merchandise called "Bag It" (29). For more information on the movie, visit: http://www.bagitmovie.com/about.html

Phthalates:

Bottled drinks also contain phthalates, which are commonly used in the U.S. to make plastics such as polyvinyl chloride (PVC) more flexible. Phthalates are also endocrine-disrupting chemicals that have been linked to a wide range of developmental and reproductive effects, including:

- Reduced sperm count
- Testicular abnormality and tumors
- Gender development issues

Despite their potential for harm, the FDA doesn't regulate phthalates or class them as a health hazard, due to the supposedly minute amounts present in plastic bottles. Unfortunately, this decision doesn't take into account the significant presence of plastics in the average American's daily life, the fact that phthalate concentration increases the longer a plastic water bottle is stored, or that a bottled drink that's exposed to heat causes accelerated leaching of harmful plastic chemicals into the drink.

Regulatory issues:

In the U.S., bottled water and tap water are regulated by different federal agencies. The Food and Drug Administration (FDA) regulates bottled water, and the Environmental Protection Agency (EPA) regulates tap water. With different policies and rules in place for each agency, the enforcement and monitoring of water quality for bottled water doesn't meet the standards set for tap water. Due to stricter EPA policies, incidents of tap water contamination must be reported immediately to U.S. citizens. However, there's no such rule for bottled water, despite numerous bottled water recalls (30) taking place over the years.

Other health concerns:
In addition to the negative impacts of BPA and phthalates on human health, there are also growing concerns regarding carcinogens and microbial contaminants that have been found in test samples of bottled water.

The Animal Impact:

Plastic bottle tops are rarely recyclable [31]. As with plastic bags, they often end up at the bottom of the ocean and in the stomachs of a variety of animal species that mistake them for food. One albatross that was recently found dead on a Hawaiian island had a stomach full of 119 bottle caps.

Marine life falls prey to this problem on a daily basis. A sperm whale was found dead on a North American beach recently with a plastic gallon bottle gumming up its small intestine. The animal's body was full of plastic material, including bottles, bottle caps and plastic bags.

The Environmental Impact:

For a quick visual of the resource impact of disposables, fill a plastic bottle with liquid so that it's 25% full. That's roughly how much oil it took to make the bottle. For a single-use disposable item, that's a lot of wasted energy.

Plastic bottles are made from a petroleum product known as polyethylene terephthalate (PET), and the chemicals used to make PET, xylene and ethylene, are extracted from crude oil. Manufacturing a quart (liter) plastic bottle requires 2.6 oz. (100ml) of crude. If you do the math for the current U.S. consumption rate of 1500 bottles a second, we are using approximately one barrel (42 gal) of crude per second, 60 barrels a minute, 3,600 barrels an hour, and 86,400 barrels a day just for plastic bottles that cause so much harm. This doesn't count the natural gas, and fuel involved in the manufacturing process and transportation to get them to you.

The other main ingredient in PET manufacturing is natural gas. Making a quart bottle requires 1.5 cubic feet (42 liters) of natural gas. That's a sphere 53 inches (135 cm) wide.

Plus, it's harder to recycle plastic bottles than you think. Of the mass numbers of plastic bottles consumed throughout the world, most of them aren't recycled [32] because only certain types of plastic bottles can be recycled in certain municipalities. These plastic bottles that aren't recycled either end up lying stagnant in landfills, leaching dangerous chemicals into the ground, or infiltrating our streets as litter. They're found on sidewalks, in parks, front yards and rivers, and even if you chop them into tiny pieces, they still take more than a human lifetime to decompose.

And, the situation gets even worse when you look at water consumption. In the case of bottled water, the plastic-making process requires over two gallons of water for the purification process of every gallon of water bottled.

So get those BPA-free bottles for your home, car, and to take along while you're walking, jogging or working out or just sitting around, and start saving the planet.

6
Paper

Got to save those trees as they are so critical to our healthy environment. Every year, Americans use more than 90 million tons of paper and paperboard. That's an average of 700 pounds of paper products per person per year consumed to produce, among other things, more than 2 billion books, 350 million magazines, and 24 billion newspapers—most of which could've been read in digital formats. And the impact of all those felled trees is compounded by wasteful disposal of an easily reusable resource. If only all of our newspapers were recycled, we could save about 250,000,000 trees each year.

Worldwide, 40% of commercially cut timber is used for paper. Evolving into a computer-based society was expected to eliminate most of our paper use. Sadly, that optimistic prediction has not yet come true; for a litany of technical reasons, we still use almost as much or more paper we did twenty years ago. Developments like cheaper printing, junk mail, and the prevalence of home printers—added to ingrained habits and the tactile psychological comfort of having a "real" hard copy in our hands—have negated anticipated gains.

Fortunately, reversing this trend is not only simple, but can make you a few bucks, too.

WHAT YOU CAN DO:

I'm as guilty as most people of falling prey to the tactile issue; I still enjoy the feel of a good book in my hands and, frankly, many technical references don't lend themselves to electronic viewing. But being a guy whose livelihood requires both an extensive library of reference material and an understanding of environmental impact, I try to offset my consumption by recycling and switching as many paper resources as possible to digital formats. While we all can't eliminate all our paper use, for whatever reasons, we can make significant progress by moving toward as many sustainable living practices as our lives and work allow.

So, in that spirit, the request I make of you is to pick at least two or three of these paper-saving solutions to start building new sustainable habits, and really go for it:

- Recycle your paper. When you start collecting all your paper, you'll be surprised at how much you actually use.
- Don't stop at just recycling your regular paper.
    a) Most cardboard packing and shipping boxes can be recycled, too; just remove the tape and paper labels.
    b) And, don't throw out all those paperboard cereal boxes and other branded packaging, either; peel off any paper stickers and tape, and keep them aside for recycling, too.
- Pay bills online. Though some folks might argue a security issue, there are now robust tools you can employ to insure your data is protected.
- Have all bill statements sent electronically, wherever possible.
- For whatever paper you can't avoid, get a paper shredder and use it. This practice not only helps in recycling, but protects your personal information, too.
- Get your family members an iPad, Kindle, or similar tablet device to download all your books online in digital format, like this one.
- If your budget doesn't allow for personal electronics, use your public library, which eliminates waste by allowing many people to read one physical copy.
- If you have to buy books, don't throw them out when you're done. Donate them to a charitable organization

or library near you.
- If you like to read newspapers and magazines, subscribe to them digitally instead.
- When taking notes, use an iPad, tablet device, or laptop to get rid of all those wasteful notepads. As an upside, electronic data can be safely kept forever.
- Instead of cluttering your life and landfills with little notes, shopping lists and post-it tags, use the list, note-taking and reminder apps on your smart phone.
- Opt out of junk mail where you can.

Now you have the solutions, go for it and make sure to get your kids involved.

Recycle for cash?
In many areas, you can take collected paper to a recycler yourself and get paid for it, but it takes a lot. Getting paid for recycling paper is not as lucrative as metal cans. You can expect to get anywhere from $50 to $75 per ton for mixed scrap paper. That's a lot of paper, so if you plan to store it in the garage, take steps to protect it from being a fire hazard.

To calculate potential recycling profits:

A good rough number to use is 500 sheets per 5 pounds:

- Note that your 500 sheets could weigh more if it includes magazine print with heavier bond paper.
- So, one ton at 2,000 pounds divided by 5 pounds for each 500 sheets means you would need 200,000 sheets to get to a ton.

Honestly, without an organized plan of collecting used paper from family, friends and neighbors, it's not probable that one household could collect a whole ton over the course of a year. But, while it might be tough to get to that

$50-$75 figure, there's still the potential of a nice reward for you or your kids in taking a few spare minutes to collect and recycle your paper. After you've turned in your recycling, treat yourselves to a movie or pizza night, or pick up a cool new toy your kid's had their eye on.

When you consider all the trees you'll save, it's a win all around.

WHY IT MATTERS:

Below are some alarming facts about our collective paper use and waste, followed by positive impacts of recycling, from research data compiled by the University Of Southern Indiana, A Tree for Each American, American Forest & Paper Association and Waste-Free Mail.com.

Newspapers:

- To produce each week's Sunday newspapers, 500,000 trees must be cut down.
- Recycling a single run of the Sunday New York Times would save 75,000 trees.
- If all our newspaper was recycled, we could save about 250,000,000 trees every year!

Paper bags:

- A 15-year-old tree can produce around 700 paper grocery bags.
- A busy supermarket could use all 700 in under an hour.
- In one year, a single supermarket can go through over 6 million paper bags!

- Imagine that already-huge number multiplied by the number of supermarkets in the United States, alone; this is why I'm so determined to get more people to start carrying cloth shopping bags instead.
- Consumer waste
- The average American uses seven trees per year in paper, wood, and other products made from trees. This amounts to about 2,000,000,000 trees per year — two trillion, every year. By trimming just 28% of that paper consumption, every American could save two trees each year.
  a) Approximately 1 billion trees worth of paper are thrown away every year in the U.S., alone.
  b) In the U.S., over 40% of municipal solid waste is paper, about 71.8 million tons each year.
- The amount of wood and paper we throw away each year is enough to heat 50,000,000 homes for 20 years.
- The average household throws away 13,000 separate pieces of paper each year. Most is easily recyclable packaging and junk mail.
- Energy and resources
- Only about one-third of the fiber used to make paper in the U.S. is from whole trees, which the industry calls round wood. It's not considered economical to use large logs for paper when they could instead be used for lumber. For this reason, only trees smaller than 8 inches in diameter, or trees not suitable for solid wood products, are typically harvested for papermaking.
- It's impossible to specify how much paper can be made from one tree, due to its complicated process and multiple factors which impact production. So, instead, calculations are based on a more consistent measure. A cord of hardwood is approximately 8 feet wide, 4 feet deep, and 4 feet high. It has been estimated that one cord of wood will yield one of these approximate

quantities of products:
- a) 1 - 2,000 pounds of paper (depending on the process)
- b) 942 one-hundred-page, hard-cover books
- c) 61,370 No. 10 business envelopes
- d) 4,384,000 commemorative-sized postage stamps
- e) 460,000 personal checks
- f) 1,200 copies of National Geographic
- g) 2,700 copies of an average daily newspaper

Benefits of recycling:

- In 1993, U.S. paper recovery saved more than 90,000,000 cubic yards of landfill space. So, the data shows recycling can work to significantly alleviate other environmental issues.
- Each ton (2000 pounds) of recycled paper can save:
  - a) 17 trees (which can absorb a total of 250 pounds of carbon dioxide from the air each year)
  - b) 380 gallons of oil
  - c) 3 cubic yards of landfill space
  - d) 4000 kilowatts of energy
  - e) 7000 gallons of water.
  - f) These figures represent a 64% energy savings, a 58% water savings, and 60 pounds less of air pollution!
- Burning that same ton of paper would have created 1500 pounds of carbon dioxide.
- The construction costs of a paper mill designed to use
- The construction costs of a paper mill designed to use waste paper is 50 to 80% less than the cost of a mill using new pulp.

7
Styrofoam

Most of the items listed in this rogues' gallery of environmental harm were once considered miracle conveniences of the modern age. Their use grew exponentially, making them ever-present time bombs so ingrained in our everyday lives, eliminating them is an uphill battle, even in the face of ironclad proof of their dangers to our health and environment. Styrofoam is no exception.

While we're finally making strides in tackling the mountains of disposable cups, to-go containers and packing materials, we still have work left to do. Though sometimes the decision to use Styrofoam is out of our hands, whether or not to take it can be. It may require a little preparation, but that extra moment of forethought can make a huge difference in the long run.

WHAT YOU CAN DO:

The solutions I'm offering are mostly simple and don't require much time or energy to do, just the willingness to be ready to say no to a Styrofoam option. Not too rough, right? And, in the cases you can't avoid Styrofoam, your input might help turn a head or two on the right path.

At work:

- The easiest solution: instead of using disposables, bring a mug to work.
    a) If you drink 3 cups of coffee or tea a day, your typical consumption of 12 oz. cups per year at 15 per week with 48 weeks at work (52 less vacation and holidays), that's 720 fewer cups used, just by you.

- Don't stop there. Why not encourage fellow employees, or even your whole workplace, to make the switch to a more earth-friendly break room?
    a) If your office or workplace has 20 people who also drink coffee or tea, you could easily be using 15,000 cups per year.
    b) 12 oz. cups typically cost about $55.00 per thousand, costing the company at least $825.00 per year. If helping the environment doesn't sell the issue, that savings could pay for one heck of an office party for 20 people.
- Encourage your workplace to stock the break room with eco-friendly mugs.
    a) Cost of the new eco-friendly mugs might run only about $4.00 per cup.
    b) As a bonus, the new mugs are a one-time investment, versus wasting money year on year for something designed to wind up in the trash.
    c) Plus, the mugs can be produced as branded promotional items.
    d) Even better, promotional mugs can be written off as a tax deduction.
    e) If people are worried about using someone else's cup, mark each with an employee's name and store on a rack in the break area.
- If your workplace has frequent guests, and mugs aren't a feasible option, switch to compostable food service versions (see Alternatives to Styrofoam below), instead. They're readily available from a host of suppliers, affordable, and the eco-friendly move is a great green image builder.

Food and drinks to go:

- Carry a travel coffee and/or drink cup with you and ask that it be used when you order drinks to go instead of a disposable.
- If you go for take-out food often, ask whether they can use your own travel containers or compostable versions.
- If your local shops still use Styrofoam cups, food service containers or packaging, encourage the managers to switch to more eco-friendly compostable versions (see Alternatives to Styrofoam below). If Dunkin Donuts could break their Styrofoam cup habit, anyone can!

At home:

- When you're mailing packages or moving, choose a more environmentally friendly option.
  a) Select packing peanuts made of compostable material (see Alternatives to Styrofoam below).
  b) When moving, skip the Styrofoam packing and use blankets, towels, and clothing in boxes of breakables to pad the base and help fill the gaps.
- Use recyclable packing paper or newspaper to wrap your breakables.
- When you can't avoid taking Styrofoam, don't throw it out.
  a) Find out if your city has a Styrofoam recycling program, and if not, encourage them to start one.
  b) Keep Styrofoam packing materials and peanuts aside to use when you or a friend need shipping or moving supplies.
  c) Reuse or repurpose it whenever possible.

WHY IT MATTERS:

We all know the selling points of Styrofoam for food service and shipping convenience. Now, here's the ugly side of the issue and why we need to break our Styrofoam habits, for mother earth and people exposed to the manufacturing process.

Styrofoam is actually the trade name for polystyrene, the petroleum-based plastic from which it's made. It's popular because of its low weight, good insulation properties, and advantage as a packing material for shipping without adding costly extra pounds. Unfortunately, for all of Styrofoam's good points, data has shown that Styrofoam also has harmful effects.

Effects on Human Health:

Elements used to make Styrofoam:
Styrene is the foundational ingredient used to make polystyrene. It's broadly used in the manufacture of plastics, resins and rubber. The U.S. Environmental Protection Agency (EPA) [33] and the International Agency for Research on Cancer [34] have established styrene as a possible human carcinogen. [35] It's estimated that 90,000 workers are exposed to Styrene every year.

Those who work in styrene product manufacturing and are regularly exposed to high levels of styrene have experienced acute health effects, including:

- Irritation of the skin
- Irritation of the eyes
- Irritation of the upper respiratory tract
- Gastrointestinal effects

Chronic exposure to styrene leads to further complications, including effects on the nervous system. Symptoms of chronic exposure include:

- Depression

- Headache
- Fatigue
- Weakness

Other Effects:

Styrofoam containers are commonly used for take-out food, but chemicals can leach into it and contaminate that food, affecting human health, kidney function and reproductive systems. This effect is further accentuated if food is reheated while still in the container. NEVER heat Styrofoam: always remove food to a cooking vessel for reheating. Microwaving Styrofoam causes the release of toxic chemicals, which poses a threat to human health.

Impacts on the Environment:

Pollution:

Styrofoam is non-biodegradable and appears to last forever. It's resistant to photolysis, the breaking down of materials by photons originating from light. This characteristic, combined with the fact that Styrofoam floats, has resulted in large amounts of polystyrene that have accumulated along coastlines and waterways around the world—it's considered a main component of marine debris. When thrown away as trash, polystyrene cannot biodegrade or breakdown via other means, remaining in the environment for thousands of years. Keep in mind, plastics cover 25-30% of space in landfills.

Foam polystyrene has been found in water and wind, especially at shores, making up for a considerable amount of marine debris. This also affects animals in the wild, due to broken down bits of polystyrene obstructing their airways, contaminating their resources, and causing cancer and digestive problems.

Styrofoam can be recycled, but the market for recycled Styrofoam is diminishing. Many recycling companies no longer will accept polystyrene products, like my recycler where I live, and it is sad. Polystyrene products that are recycled can be remanufactured into things like cafeteria trays or packing filler.

Air pollution:

Along with the health risks associated with the manufacture of polystyrene, air pollution is another concern. The National Bureau of Standards Center for Fire Research has found 57 chemical byproducts released during the creation of Styrofoam. Polystyrene is manufactured with HCFC-22, a greenhouse gas that affects the ozone layer. The polystyrene manufacturing process is the 5th largest creator of hazardous waste. The process of making polystyrene pollutes the air and creates large amounts of liquid and solid waste.

Climate change:

Styrofoam manufacture uses hydro fluorocarbons (HFCs), which negatively impact the ozone layer and climate change. HFCs are less detrimental to the ozone than chlorofluorocarbons (CFCs), which were used in the manufacturing of Styrofoam in the past, but it's thought that the impact of HFCs on climate change is much more serious. Styrofoam is also made from petroleum, which is a non-sustainable resource, the production of which creates heavy pollution and accelerates climate change.

Alternatives to Styrofoam:

Compostable food service packaging is very trendy right now as an ecologically correct option. Compostable containers are made using cornstarch, palm fiber, peat fiber and wheat stocks, and they're able to break down into soil-enriching compost.

Think getting your favorite places to switch isn't possible? You'll never know until you try. Dunkin Donuts actually stopped using Styrofoam cups! If we can get lots more companies doing it, and even press our local and state governments to ban them for alternatives, the world can be really impacted.

Long-term solutions:

So now you know the bad news about Styrofoam, but have a solution to rid our lives of this nasty stuff, even if you just do it on your own. But maybe you can convince your boss to take the impact of your changes wider. Or, maybe you are the boss, so have the opportunity to impress your staff and clients with your new greener attitude, and make them proud that your company cares.

Want to make your impact even bigger? If you're the letter writing type, this is a critical issue to write your local, county, and state governments about. Some of America's biggest cities, as well as entire counties and states, have already made the move to completely or partially ban Styrofoam. Why not yours?

Green and sustainable practices aren't just fringe ideas anymore. Positive environmental activism and public outcry get results. Here's a list of state, county, and local governments that have already completely or partially banned Styrofoam, compiled by Groundswell. (36)

- New York City (and several other cities in New York)
- Takoma Park, MD
- Seattle, Washington
- Washington DC
- Miami Beach, FL
- Freeport, Maine
- Portland, Maine
- Nantucket (City & County), Massachusetts
- Minneapolis, Minnesota

- Portland, Oregon (and several other Oregon cities)
- Los Angeles County and San Francisco, California (and many other cities and counties in CA)

8
Litter

You may not think litter is a sustainability issue but it is because the careless disposal of trash impacts flora and fauna in a big way and causes the society to expend all sorts of resources to deal with your choice to pollute the environment.

The issue of litter bothers me far more than many others, which is probably to be expected. Most people concerned about the environment are irritated by self-involved people who carelessly chuck stuff out a window onto the road or in a parking lot to get it out of their immediate world. Problem is, of course, they've punted that trash into all of ours, infringing on everyone and everything else that lives in our habitat and the larger biosphere.

I often wonder, when I see someone intentionally litter, how anyone could have an attitude so uncaring and obnoxious that they think it is okay to dump their trash out into the environment. Is it a power trip, making us clean up their crap? Or maybe they just don't care? It's not like garbage cans don't exist everywhere. Littering has always struck me as one of the most stupid things earth dwellers do in the U.S. and other developed countries, and has bothered me since I was old enough to understand the damage.

If this extended rant sounds judgmental, you're absolutely right; because, when things we do as people are just plain wrong, another inhabitant on this planet has every right to call it out. What follows is ways to recycle that irritation into a positive impact for the earth.

WHAT YOU CAN DO:

Be more aware of the problem:

You're already a step ahead right now. My hope is that everyone reading this book who litters, even occasionally, will stop. More importantly, I hope you make sure the kids in your life are aware that this is bad on so many levels that impact us all. As a reality check, I understand that many underdeveloped countries lack any real functioning waste management systems, and it's a problem that contributes to disease and pestilence with no easy fix.

But, in the vast majority of places, littering is a choice.

- Research shows 85% of littering behavior is the result of individual attitudes.
- 75% of people have admitted to littering in the past 5 years.

Choose not to litter:

How?

- Keep a trash bag in your car, boat, camping and beach supplies, and hiking gear for any place you go that doesn't have a readily available trashcan AND USE IT.
- Don't let people who are part of your group or family litter. Don't just let it slide. Say something, and maybe even share one or two of the facts below.
- Talk with your kids about developing the positive habit of never littering and speaking up when they see it happen, and be their good example to follow.

Be a part of the solution:

One great idea for you and the kids is to volunteer to do a litter cleanup event in your community. Participating in an active cleanup will not only make you feel good about helping the planet, but for kids it also drives home the real extent of the problem.

And, if there aren't yet existing cleanup efforts in your community, why not start one? Or, why stop at just one? Get different groups involved, from your neighborhood, your clubs, your church, and even your kids' school and teams. Let Other people who can't take part support your efforts by sponsoring a participant, and use the funds raised for charity or other environmental improvements to your community.

When you set your mind to it, you'll come up with countless ways to make a difference.

WHY IT MATTERS

Some facts about the damage:

- The impact of littering causes harm to the environment, pets, and wildlife. Birds, fish and other ocean-dwelling animals are often unable to distinguish between trash and food. As a result, 1 million sea birds and 100,000 marine mammals die each year after becoming entangled in or ingesting litter.
- Altogether, our collective disregard adds up to around 52 billion pieces of litter cluttering up the landscape. That breaks down to more than 6,700 items per mile.
- Littering doesn't just take a toll on environmental quality; there's also an economic cost. Between cleanup efforts, decreases in property values and medical expenses associated with treating illnesses caused by litter, the cost adds up to $11.5 billion each year.
- Litter can also clog storm-water drains and cause flooding.
- The most common forms of litter are food/organic material, cigarette butts, and small pieces of paper — receipts, gum wrappers.

- Food scraps and other organic items that are disposed of improperly can increase algal blooms in water, which reduces the amount of available oxygen for other aquatic life, such as fish.
- Litter can build up and attract insects and rodents, which bring unwanted germs and disease to the ecosystems.
- If trash is sitting in water, the water becomes contaminated, and when the water evaporates whatever was in the trash in now in the air.
- According to a study by Green Eco Services (37) about 25,000 accidents are caused in the US by litter each year.

The extent of the problem:

Despite being bombarded with media campaigns that warn about the dangers of littering, many Americans aren't heeding the message. A willingness to litter crosses all demographic and class lines, and is more prevalent than you might think.

Here are a few facts to illustrate the extent of our litter problem:

- 75% of people have admitted to littering in the past 5 years.
- Following are some figures from the Don't Mess with Texas program (DMWT) literature. I used their numbers because I live in Texas, but the rest of states where I could find data are as guilty, with similar comparisons by age groups.
  a) 1 in 4 Texans admits to littering in the past year.
  b) DMWT asked how many times respondents had littered in the past month. Breaking littering down by age groups:

i. 16–24 years old = 68%
   ii. 25–29 year-olds = 60%
   iii. 30–49 year-olds = 50%
   iv. 50 years of age or older = 33%
- DMWT also tracked the same question in respondents 16–29 years old by their residential environment:
   a) In urban/suburban counties = 69%
   b) In rural counties = 58%
- DMWT also looked at whether we teach our children not to litter:
   a) Nine in ten parents surveyed have told their children not to litter, and 97% would pay more attention to littering if their children asked them to.
   b) But only one third (32%) of parents say their child has talked to them about littering (45% among parents who are 16–29 years old and would therefore have younger children).

9
Wood and Wood Furniture

Wood preservation in many forms are all major sustainability issues, and next to paper usage are the most impactful. The preservation issue combines usage like paper and wood furniture and wood products as a consumer impact but also from a manufacture standpoint using sustainability sourced materials is where a major positive impact starts and will be discussed in detail later in the book. But there's one other aspect of what we do with wood we have that makes my list of most irritating problems. We waste a lot of wood and don't seem to realize the senseless harm we're doing to our environment.

Eventually, the wood we have in our homes or work always seems to find its way to the dump, when we could keep it for a variety of uses instead. Wood never spoils or goes bad and, if stored with just a little care, it'll be there whenever you need it. Even if you can't find a new use for your project leftovers, remodeling demo scrap or piece of wood furniture you don't like or need any longer, odds are someone else can.

WHAT YOU CAN DO:

- Measure twice, buy once! Plan and organize your space ahead of time before purchasing furniture pieces.
- Try multiple-function furniture pieces like convertible sofas (38) and futons (39). Having a guest bed and a sofa built into one piece of furniture helps save on important materials such as wood and reduces deforestation.
- Recycling isn't just for paper and cans!
    a) Donate your old furniture to the Salvation Army or Goodwill, or sell what you no longer need so it can be reused instead of discarded.

- b) Get your new look from a vintage dealer, furniture reseller, or charitable resale organization such Salvation Army or Goodwill.
- Use slipcovers (40) to keep your furniture looking fresh and new for years to come.
- Renew your chair by removing the seat and fabric — check to see if the foam is still usable (no mold) — and replace the fabric with new (preferably organic) fabric with a staple gun.
- When remodeling seems to call for new furniture, try asking the following questions, and perhaps you can get the redesigned look you want and save some money without actually buying new stuff:
  a) Do you really need new furniture?
     i. Sometimes your existing furniture just needs a little shifting around, new covers, or a few accessories to create a fresh look for your room décor.
  b) Could readjusting your existing furniture make your rooms look new?
     i. Try switching furniture pieces from different rooms. A simple dining room table can be used as a desk, or a bookshelf can be used with storage bins to store socks and clothes. A little creativity can go a long way.

- Check online for projects for spare wood uses. You'll be surprised what great stuff you'll find and, believe me, being a carpenter is not required.
  a) A suggested site is Pretty Handy Girl. Great projects are available at
     https://www.prettyhandygirl.com/practically-free-scrap-wood-projects/

WHY IT MATTERS:

Disposal of wood and wood furniture not only unnecessarily consumes valuable resources, but wastes critical space in landfills.

- In 2009, the U.S. EPA reported that furniture accounted for 9.8 million tons (4.1%) of household waste.[41]
- Furniture is the number one least-recycled item in a household, and it was only up until 2008 that there was a recovery (materials used for recycling) greater than 0.05% since the 1960s.
- Remodeling, that often results in wood furniture being discarded, contributes to a 2015 estimated $121.7 billion [42] spent to refurbish our homes with new furniture.
- EPA numbers for 2015 show wood generated as 16.30 millions of tons: 2.66 millions of tons recycled, 2.58 million tons combusted for energy recovered, and the rest — over 11 million tons — landfilled.
- United States Department of Agriculture (USDA) Forest Service estimates that 20% of the total volume of wood used for construction ends up as waste.

Before you send that leftover wood or unwanted furniture to the dump, please think about the tree or trees — not to mention the critical landfill space — you can save by finding a better solution. Just a little effort will not only make you some money, but help the environment, too.

10
Buy a Rain Barrel

This activity compliments the reduction of water usage discussed earlier and adds a real way to increase your available water instead of reducing consumption. The EPA estimates the average American family uses 300 gallons of water per day, enough water to fill a six-person hot tub. According to the nonprofit organization Circle of Blue, that water could cost you more than $50 per month (depending on where you live), and the price of water is on the rise. In fact, it rose 41% between 2010 and 2015.

Plus, the impact goes beyond the environment or your budget. In an emergency situation, it can be extremely valuable to have 40-60 gallons or more of water available. Particularly for those who live in isolated, rural areas or areas prone to wildfires, storing rainwater can offer extra protection and peace of mind.

WHAT YOU CAN DO:

Selecting a rain collection system:
The first step is to invest in a rain barrel or other collection system.

Rain Barrels:
Rain barrels come in various sizes from 40-60 gallons, and most allow you to connect two or more together to collect even more water.

Roof runoff:

Below is my rain barrels and are 60 gallons each and used for gardening and for any emergencies as indicated by the down spot are supplied by roof runoff.

They need to be positioned to receive water and some adjustments to your roof gutters, but that's all pretty easy.

Collection systems that use roof runoff have a few cautions to be aware of (noted below), but for most concerns, there are very economical filters that can be added to your system. The exceptions would be issues regarding lead, bitumen, tar or treated timbers.

Free-standing:
If you have potential issues with roof runoff collection (noted below), like lead or bitumen, or don't want the chance of potential contaminants, it's possible to use a collection system that doesn't use your roof runoff as a source. You can build a collection funnel like the one below from Pinterest.

From - https://www.pinterest.com

Underground tank system:

If you're really motivated to put in the most efficient rainwater collection system, you can also install one that's comprised of big tanks under the ground. When it rains, the water is directed to the tanks. Then, when the rain water needs to be used, there's a filter and pump that sends the water out for your needs. Of course, this type of collection system is much pricier, and in most cases will require installation by a professional, but I've seen systems installed by a couple of people who were highly motivated to make it work.

How it works:

Simply placing a tap on a rainwater tank can supply rainwater for outdoor uses such as garden watering, car washing, or topping-up your swimming pool or spa. Using water indoors for toilet flushing, clothes washing, bathing, showering, and in your hot water service will require additional plumbing and a pump.

If you plan to use water for drinking on a regular basis, you should have an additional filtering system or Ultra Violet Light treatment, but it really depends on the pollution in your area. This step applies only to drinking the rain water; all other uses indicated are no problem.

But, please, use common sense. If the water looks strange, have it checked or don't use it for human contact or ingestion purposes.

Some specific uses for rain water:
All the uses add up to several hundreds of gallons a year of savings in most climates.

Composting:

- Water helps break down your compost and aid in decomposition. You don't need filtered or tap water for this purpose.

Landscaping & Gardening:

- Rain is already the natural provider of water for your plants and trees, usable as long as it has not come in contact with contaminants. Collecting it will make it more accessible to be used in watering cans or indoor plants.

Pets & Animals:

- Rainwater is typically safe for animals to consume, including livestock. You can also use rainwater for birdbaths or to bathe your pets.
- Just be sure you have established a clean and safe way to collect water before giving it to animals.

Produce Rinse:

- Rainwater is also a perfect vegetable and fruit rinse. If you're a gardener, you can take the produce straight from your garden to your very own rinse bath.

Cars & Equipment:

- Next time your car needs a wash or you want to clean up your lawn mower, use rainwater. It's also perfect for cleaning garden equipment, bikes, tractors, trailers and most equipment found in your garage or shed. This is a big savings that you need to really consider maximizing, as these uses can be done no matter what kind of water you collect.
- One important note on the carwash issue. If you can't use collected rain water, to save valuable community water, take your car to a car wash instead of washing it in the driveway. They actually use a lot less water per car than you would doing it at home, and often also have a recycling system.

Water Features:

- Rather than pumping water from your hose into outdoor water features such as fountains, ponds or pools, consider using rainwater.
- For water features and pools, be sure to filter your rainwater first, to avoid clogging or excess debris.

Fire Safety:

- Particularly for those who live in isolated, rural areas or areas prone to wildfires, storing rainwater can offer extra protection and peace of mind. Install an easily accessible and functioning pump so that you can access water quickly, in case of an emergency.

Flushing Toilets:

- Did you know flushing toilets uses more than 25% of your home's water consumption?
- Place a bucket of rainwater next to your toilet, and next time you're ready to flush, pour water into the bowl. Your toilet will automatically flush with the added liquid. Your bucket should hold the same volume of liquid as your toilet tank.
- This is important, especially in emergencies when water may be turned off. Inside toilets can be easily managed with the limited water source in your rain barrel.

Washing Outside Areas:

- For cleaning windows, porches, decks or the exterior of your home, consider using rainwater as an alternative to hose water. Rainwater is considered "soft water", so there's no risk of damaging paint or windows.

Drinking and Cooking:

- Rainwater can even be used for personal consumption when safely filtered and treated!
- Consuming rainwater is only safe once it has been filtered, cleaned and tested. You should also consult local government restrictions or guidelines before consuming.

How much money can you save?

Your water meter measures cubic feet, not gallons. Each cubic foot equals 7.48 gallons. EMWD bills in 100 cubic foot increments— called "billing units"—equal to 748 gallons.

From a personal perspective, my water bill is an average of $39.13 per month in Texas for 8000 gallons. That's about 0.005 cents per gallon or $5.00 per 1000 gallons, using rain water where I'm at — which accumulates at an average of approx. 46,000 gallons per year, according to "The Texas Manual on Rainwater Harvesting" put out by the Texas Water Development Board — with the 2,500 square foot collection surface I have. (It's actually probably more, with a 2,800 sq. ft. single story house, but I don't have a contained system so am only collecting about one third of the possible water.)

If I use an average of 15,300 gallons a year, I'm saving about $77 per year. Granted, I've never measured it exactly, and just keep using the spare water until the barrel goes dry for my wife's garden, lawn watering, car washing and even some for the dog's extra bowl outside. I would obviously have a much larger bill if I didn't use the water collection barrel. And, we haven't even gotten into inside water uses yet.

However, while working on Maria hurricane support in Puerto Rico, we did use rain and cistern water for everything because it was all we had. It really does work, including the shower and toilet flushing stuff.

A few cautions:

- THIS IS IMPORTANT: Drinkable rainwater can be collected from most roof types. However, you should avoid roofs with lead flashings, covered with lead-based paints, bitumen and tar or treated timbers. Roofs with these properties will contaminate your water supply, so don't use the runoff water for any reason unless you have professional help on designing an effective filtering system.
- If your roof is newly installed or freshly painted, it's best to wash it down and discard the first few runoffs of rainwater if you're using your tank water for drinking or cooking.

- Contamination can happen from a number of things, so check whether filters should be added to your system for:
    a) Animal droppings (e.g. from birds, bats, possums, etc.)
    b) Mosquitoes and frogs getting into your rainwater supply
    c) Bacteria, parasites and microorganisms
    d) Chemical spraying in your area
    e) Air pollution from nearby industries like manufacturing plants, spray painters, chemical plants and quarries or vehicles associated with freeways and main roads

WHY IT MATTERS:

The EPA estimates the average American family uses 300 gallons of water per day. To put that number into perspective, that's enough water to fill a six-person hot tub. How much does it cost to fill this hypothetical hot tub? According to the nonprofit organization Circle of Blue, it could cost you more than $50 per month (depending on where you live).

The price of water is on the rise, too. In fact, it rose 41% between 2010 and 2015.

So, with a simple rain barrel, you can help save the planet and your water budget.

SECTION TWO: MOVING FORWARD

"There is no moral precept that does not have something inconvenient about it."
~ DENIS DIDEROT

In your journey of the process of help supporting a sustainable environment and building a more earth-friendly lifestyle, you've already taken the first step: acting on the desire to change old habits and develop more sustainable ones. Through the previous activities, I addressed a litany of issues that harm the environment and small changes you can make to lessen your negative impact. From reducing energy, paper and water waste to eliminating harmful litter, plastic bags reduction, most offered simple solutions that required little to no investment or strenuous effort.

Mainly, you've established the foundation for a new focus on mindful living.

Now, it's time to move forward. Like the simple tips in the last chapter, the actions in this chapter don't require special training or monumental effort. Most rely primarily on a willingness to invest a small amount of time, effort or materials. But, unlike the prior chapter's more one-to-one rewards of eliminating one source of harm, in this chapter you'll find longer-term actions that have the potential to increase your positive impact and monetary savings exponentially.

Sustainable living with mindfulness to environmental impact, when done right, is not a sprint but a marathon. Short-term investments and good habits yield ongoing benefits that keep accruing over time. Fortunately, most of these actions are as simple as the tips that have preceded them; they simply call for either a minor up-front investment or the discipline to establish and maintain good practices.

So, if you want to take your environmental game to the next

level but don't have a lot of time or money for major changes, you'll find a few great ideas to keep moving forward.

11
Batteries & Solar Chargers

Rechargeable batteries and solar chargers are more than just a must-have for your camping and emergency kits. When used as a primary energy source for your phones, tablets, cameras and other devices, they can really help reduce costs and environmental impact.

The fact is: people are using more and more batteries. Consequently, they're also throwing away more batteries than ever before. According to the Environmental Protection Agency (EPA), each year Americans throw away more than three billion batteries. That's about 180,000 tons of batteries taking up valuable space in landfills, including some types that still contain toxic constituents like mercury and cadmium, which can pose a potential threat to human health and the environment.

And, if the prospect of lessened environmental impact and landfill strain isn't a compelling enough reason to rethink your battery habits, a change can you can save you a lot of money, too. In this section, I'll cover the positive impact of rechargeable batteries, battery recycling, and solar chargers as well the long-term gains for your wallet.

WHAT YOU CAN DO:

Switch to rechargeable batteries:

About three billion batteries are sold annually in the U.S., averaging about thirty-two per family or ten per person. The average person owns about two button batteries, ten normal (A, AA, AAA, C, D, 9V, etc.) batteries, and throws out about eight household batteries per year. Those discarded batteries can really add up. Now, imagine the positive impact if all those people switched to rechargeable versions that can be reused hundreds of times before being recycled. That'd save our environment a lot of landfill space and resources.

Plus, single use batteries may seem cheaper, but over just a short time rechargeable batteries pay for themselves and reap sizable savings for your minimal investment. On average, rechargeable batteries cost only about $0.65 to $1.20 each for AA, $1.50 to $2.50 for AAA, $3.50 to $4.50 for C, and $1.50 to $3.00 for D. Sure, that's higher than what you'll pay for a standard battery, but that minor uptick in price is dwarfed by their sheer potential. Rechargeable batteries can typically be recharged and reused 300-500 times, conservatively, depending on the specific battery manufacturer.

That's a huge difference for such a tiny investment. How does that add up for you? I'm just getting to that. Hang on to your hats, because the numbers might blow them right off.

What's in it for you?

How would you like to get around $1,280 worth of batteries for only $61?

Here's a little simple math to break down your benefits:

- In a recent online check, Walmart listed 8 AA Energizer (alkaline single use) batteries for $6.39, about $0.80 per battery.
- In the same check, 4 Energizer AA Rechargeable batteries cost $10.95, about $2.74 each.
- A charger costs around $50, bringing the total investment for 4 rechargeable batteries to approximately $61.
- You can typically recharge batteries 300-500 times, conservatively, depending on the specific battery manufacturer.

- So, even if your four batteries slightly under-perform and provide only 400 recharges, you'd get the energizing power of 1600 single-use batteries.
- You'd have had to buy 200 of the single-use 8-battery packs to equal the power of your four humble rechargeables, at a cool cost of $1,278.
- Now, multiply that by all the batteries your family uses in a year.
- Imagine what you can do with all the money you've saved.

Bear in mind, too, these are really conservative numbers. Not only did I underestimate the number of potential charges, I didn't take into account the potential for extended use per charge of rechargeable batteries. According to Green Batteries.com, "For most high drain electronic devices, like digital cameras, rechargeable batteries will continue to work much longer than alkaline batteries. In fact, in devices like digital cameras, NiMH batteries will run on a single charge for 3-4 times as long as they would on an alkaline battery."

Yes, you read that right. *In some cases, you have the possibility of your devices running three to four times longer on a single charge on rechargeable versus alkaline batteries.*

So, think of that $1,280 as a starting point and see how much more you save.

This number doesn't even include the electricity savings if you use solar chargers.

Below are some of my rechargeable batteries and the recharger. Notice that the batteries have USB connections that not only use the charger but can go into any USB port to charge. I have about 20 of the batteries and use very few non chargeable batteries in a year as I only keep them around for the grandkids toys that will be going back home with them. Yes they do work and I have saved a lot of money.

Use solar chargers:

In recent years, solar chargers have moved from exotic gadget catalogs and science fiction to everyday life. Not only can solar panels run your home's power, but your devices and batteries, too. Want to boost your savings in the switch to rechargeables? Pick up some solar chargers to save on your electric bill and your pressure on overworked power grids.

There's a wealth of models and varieties of solar chargers available, for home or on-the-go, with some as affordable as under $20 all the way up to over $500. To maximize your investment, get one that not only charges your devices directly but will recharge your batteries as well. Here are just a couple specific ones (I don't get anything for their use just wanted to show you examples:

As Promised Some Detail on Solar Battery Chargers:

Solar charger Soluser 10000mAh Portable Solar Power Bank, IP67 Waterproof Dual USB Ports Battery Bank with 2 LED Flashlight, Compass for smart Phone, cell Phone, iPhone, Sam- sung, lg phone, Android phone. Cost approximately $30 but have seen them on sale for $20.

Solar Powered Battery Charger Charges 4 D, C, AA and AAA Batteries. Cost approximately $45 but have seen on eBay for less than $40.

Below are a few items I use a lot besides other solar charges not shown. The two solar lanterns also can be charged with USB connection. They really do work well!

Recycle your batteries:
 According to the Environmental Protection Agency (EPA), each year Americans throw away more than three billion batteries. That's about 180,000 tons of batteries. While the fact that since the 1996 federal law banning mercury in batteries, current single-use alkaline batteries are now considered safe to throw in the trash, "safe" doesn't negate the possible impact or minimize our overall resource expenditure in making them a one-time-use bunch of metals. Even if you don't switch to rechargeable batteries, it'll make a great impact to simply turn in your single-use versions to a battery collection center.

What to Do With Your used Batteries:

Even though it's "legal" to dispose of many personal use batteries, just a little effort to turn them to a collection center for recycling can go a long way to minimize your and your family's personal impact on the environment by using them. Rechargeable lithium ion batteries can be recycled, but only at specified locations. When you discard them, they must be disposed of at a household hazardous waste collection point (check with your local landfill) or battery recycling drop off location, NOT placed in the trash.

Below are some major companies' positions on taking them back for recycle, besides collection centers in your community or saving them for the Household Hazardous Waste collection events.

Save them to take a few "recycling" trips a year, with the following guidelines:
1. Store each type in its own container or plastic bag.
2. Keep batteries from different manufacturers separate.
3. Store batteries at room temperature or below, but it is not necessary to store them in the refrigerator or freezer.
4. Cover the terminals with masking tape or plastic caps (especially 9 volts).

Specific guidelines per battery type:
Raw Materials Company Inc. (RMC) has provided the following guidance you can view online at https://www.rawmaterials.com/page/education/prepare-batteries/

Most of the batteries that we use every day do not require any special preparation before recycling, however we do recommend that precautionary measures be taken when recycling certain types of batteries to reduce any risk of short-circuit.

- Primary Lithium "Button" Cell Batteries (non-rechargeable): Special consideration must be taken to ensure all primary lithium "button" cell batteries have their positive terminals taped prior to recycling. We recommend taping both the positive and negative terminals by simply placing a single piece of tape around the top and bottom of the button cell covering both terminal ends.
    a) Button cell batteries are used in many applications such as musical greeting cards, watches and hearing aids.
- Other Primary Lithium Cell Batteries (non-rechargeable): Aside from button cell sizes, there are the more common AAA, AA, C, D and 9 Volt varieties. All of these batteries must have their positive terminals taped before recycling.
- Sealed Lead Acid Batteries (SLA): Sealed Lead Acid batteries are commonly used to power emergency lighting systems, UPS power units, remote control cars and vehicles. The sizes of these batteries vary depending on the application and each must have their positive terminals taped before recycling.
- 6 Volt Batteries: 6 Volt batteries are the type used in larger flashlights and lanterns. Although not subject to the Transportation of Dangerous Goods Act, we recommend that the protective caps or masking tape be placed over its terminals prior to being recycled as well.
- All 9 Volt (Including Alkaline): 9 Volt batteries are the type usually found in smoke detectors and alarm clocks. Simply place a piece of masking tape over the terminal ends to secure both the positive and negative terminals.

- Rechargeable batteries: All rechargeable batteries need to be recycled, even though in some situations you can put them in the trash. To avoid any confusion, just recycle them all. These major companies provide convenient rechargeable recycling options for their customers:
    a) Lowe's: Lowe's says that the recycling centers will offer a convenient and free way for customers to recycle rechargeable batteries.
    b) Best Buy: Customers can drop off rechargeable batteries at any U.S. and Puerto Rico Best Buy stores. Best Buy does not accept alkaline batteries for disposal. Customers should call 1-800-RECYCLING or visit www.1800recycling.com to find alkaline battery recycling centers in their neighborhoods.
    c) Home Depot: Bring in rechargeable batteries where they'll recycle these items for you at no charge.
    d) Staples: Bring in your rechargeable batteries that no longer hold a charge to your local Staples® store and they'll recycle them for free. Collected batteries are then sorted and recycled into new batteries and devices.

Where can you recycle your batteries?

If you're looking to discover where to dispose of batteries, helpful resources are available online at:

- Earth911 (https://earth911.com/) has an accurate Recycling Locator for all types of batteries where you enter your ZIP code to find the nearest battery recycling center
- Call2Recycle (https://www.call2recycle.org/) offers recycling resources

- RecycleNation (www.1800recycling.com) provides a resource to find alkaline battery recycling centers in your local area, or you can call 1-800-RECYCLING.

WHY IT MATTERS:

Duracell's website says: "Alkaline batteries can be safely disposed of with normal household waste." Energizer confirms that regular batteries are fine to toss in the trash, but says rechargeable batteries should be recycled according to US federal guidelines.

But the fact that you can put them in the trash doesn't mean it's good, just easier.

When you check various web sites, you'll find a conflict in the position that lithium ion batteries can or cannot be thrown in the trash. Per American Disposal Services "It is also important to note that Lithium Ion batteries cannot be disposed of in your trash either! ... When you discard them, they must be disposed of at a household hazardous waste collection point (check with your local landfill) or battery recycling drop off location, NOT placed in the trash." What's clear, by regulation, is that they aren't considered hazardous waste but are subject to other requirements such as Universal Waste rules and/or specific landfill permits.

The bottom line to be compliant and insure positive environmental outcomes you should always recycle them.

Battery recycling benefits:

According to the EPA, these are some benefits of recycling:
- Reduction in waste sent to landfills
- Conservation of natural resources, such as metals and minerals
- Helps prevent pollution by reducing the need to collect new, raw materials
- Saves energy
- Reduces greenhouse gas emissions that contribute to global climate change

- Helps sustain the environment for future generations
- Helps create new, well-paying jobs in the recycling and manufacturing industries in the United States
- Once the materials are recycled they can be reused in making new products
- Additionally, keeping heavy metals out of the landfill prevents these metals from reacting to rain water that seeps into the landfill forming a toxic soup, called leachate, that can get into the ground water contaminate the supply.

What if we don't recycle?

Expanding our thinking for a moment, let's explore the global benefits of recycling, since we only have one planet and need to share and conserve its resources. According to the Global Footprint Network, if everyone in the world consumed as much as the average American, we would need four Earths to sustain us all.

There's a growing middle class around the world, and they'll want the modern amenities people in developed countries enjoy, including electronics powered by batteries. Well, there's a finite amount of natural resources and space to build landfills to support the 7 billion people that occupy earth. By 2030, that number's expected to increase to 9 billion with 3 billion new middle-class consumers. Running the math on 3 billion new battery users—using America's rate of 3 billion batteries thrown away annually—that factors out to nearly 28 billion batteries thrown away annually.

Recycling benefits everyone. It enables the reuse of materials that would otherwise be used and discarded. Recycling promotes the "take, make and reuse" circular model that will help to sustain our planet for generations to come. Turning away from the linear economic model of "take, make, toss," which is not going to carry all of us into a thriving or sustainable future.

12
Light Bulbs

This next activity is about small but important things that can greatly reduce your overall energy costs. And, yes, they're all about your light bulbs. It's easy to overlook something as ever-present and seemingly passive. But Americans spend up to 25% of their electrical utility bill on lighting.

Want to reduce your lighting costs and your drain on the electrical grid? Follow a few simple tips and, with the right bulbs, you can.

WHAT YOU CAN DO:

Turn off your lights:
Turning off lights in rooms you aren't using seems like a no brainer. But if you have a house of more than 3 rooms, it's not done as often as most folks think, if you check your actual activity over a couple days. It's a simple monitoring event you can do yourself, and I recommend not telling the rest of the household. If you let everyone in on your plans, the audit will likely be biased, because they'll make the effort to turn things off, knowing you're tracking their use.

- First, read through the information below to determine the types of lighting sources you have.
- Then monitor your habits every half hour for two days when you're home, to see what lights are left on.
- You'll be able to calculate the savings (see section below) by just cutting down the hours when lights are on in empty rooms.

The guidelines from the Department of Energy (DOE) below, with some of my comments added, should help determine your savings for reducing electricity use.

Switch to LED:

Before I move on to potential cost savings, it's important to note that overall there's no better option than LED. While other energy-efficient choices may actually conserve energy better, their other environmental impacts negate the positives. And LED lighting can be turned on and off with no impact on bulb life, so are ideal for most uses, especially combined with occupancy and daylight sensors.

Plus, a 6-watt LED bulb gives the same intensity of light as a 50-60 watt incandescent bulb. They may cost a little more, up front, but more than make up for that in long-term savings. There's really no down-side. (By the way, of the LED light bulb brands, CREE is my favorite.)

How much can I save by switching to LED?

The short answer? A lot.

LEDs have been around for many years, but only lately have manufacturers started to expand the product line from the simple diode lights that were the benchmark a few years back to newer and sophisticated versions. Brighter and more intense LEDs has been incorporated into flashlights, outdoor solar lights, and head lamps. And, only recently, new home-based LEDs have hit the market. What about the cost? Yes, they're not the cheapest bulbs available. But there's so much more to this story than just the initial cost of this bulb. You'll be surprised at how much money you can save by switching to this energy efficient lighting.

Here, I've broken down the total cost of operating 10 bulbs over 30000 hours, including electricity at .12 USD/kWh and the initial bulbs plus any necessary replacements:

- Using incandescent bulbs will cost you $2389.23
- LED bulbs will cost $499.65
- You save $1889.58

Not bad for just screwing in a few light bulbs.

Determine your lighting type:

The cost effectiveness of when to turn off lights depends on the type of lights and the price of electricity. Many people don't always know what type of lighting they really have; you may need to do some checking of what is actually in the fixture. Maybe you have some replacement bulbs you could check instead of the fixture. If not, just make sure the light has been off for a few minutes, so it's not hot to touch when you take it out to check.

An important note: The type of light bulb you use is important for several reasons. All light bulbs have a nominal or rated operating life, which is affected by how many times they're turned on and off. The more often they are switched on and off, the lower their operating life.

Incandescent lighting:

An incandescent light bulb, incandescent lamp or incandescent light globe is an electric light with a wire filament heated to such a high temperature that it glows with visible light (incandescence).

Incandescent lights (52) should be turned off whenever they are not needed, because they are the least efficient type of lighting. 90% of the energy they use is given off as heat, and only about 10% results in light. So, turning lights off will also keep a room cooler, an extra benefit in summer months. This type of bulb is the most inefficient but cheapest on a per-bulb basis, though not long-term versus more expensive and energy-efficient bulbs.

Incandescent lighting should be your prime target for replacement.

While they were once thought of as a miracle invention, it's time to move on, since they're the most energy-wasting lighting. Just plug in your home data at the web site provided later on in this section, and you'll quickly see your savings add up.

HALOGEN LIGHTING:

A halogen lamp — also known as a tungsten halogen, quartz-halogen or quartz iodine lamp — is an incandescent lamp consisting of a tungsten filament sealed into a compact transparent envelope that's filled with a mixture of an inert gas and a small amount of a halogen, such as iodine or bromine.

While halogens (53) are more efficient than traditional incandescent bulbs, they use the same technology and are far less efficient than compact fluorescent lamp (CFL) and light-emitting diode (LED).

It's best to turn these lights off whenever they're not needed.

CFL LIGHTING:

A compact fluorescent lamp (CFL) — also called compact fluorescent light, energy-saving light, and compact fluorescent tube — is a fluorescent lamp designed to replace an incandescent light bulb; some types fit into light fixtures designed for incandescent bulbs.

Why I don't recommend CFLs:

Even though CFLs are much more efficient than most lighting, they also contain mercury, so I believe the promotion of their use is misguided. It's one of those unintended consequences of earlier technological development. The tradeoff for cost savings we create, in my opinion, isn't worth the more significant problem of mercury in our environment. I hope, instead, you'll consider LED (below) as your light source whenever possible.

What if I already have CFL lighting?

Since they're already very efficient, the cost effectiveness of turning CFLs (54) off to conserve energy is a bit more complicated. A general rule of thumb is:

- If you'll be out of a room for 15 minutes or less, leave it on.
- If you'll be out of a room for more than 15 minutes, turn it off.

The operating life of CFLs is more affected by the number of times they're switched on and off. You can generally extend the life of a CFL bulb more by switching it on and off less frequently than if you simply use it less. And, considering the mercury content of CFL bulbs, replacing them less frequently is also a benefit for the environment.

It's a popularly held belief that CFLs use a lot of energy to get started, and its better not to turn them off for short periods. The amount of energy varies between manufacturers and models; however, ENERGY STAR© rated bulbs are required to endure rapid cycling for five-minute intervals to ensure they can hold up to frequent switching.

In any case, the relatively higher "inrush" current required lasts for half a cycle, or 1/120th of a second. The amount of electricity consumed to supply the inrush current is equal to a few seconds or less of normal light operation. *Turning off fluorescent lights for more than 5 seconds will save more energy than will be consumed in turning them back on again.* Therefore, the real issue is the value of the electricity saved by turning the light off relative to the cost of changing a light bulb. (55) This, in turn, determines the shortest cost-effective period for turning off a fluorescent light.

The Value of the Energy Saved by Turning a CFL Off Depends on Several Factors:

The price an electric utility charges its customers depends on the customer classes, which are typically residential, commercial, and industrial. There can be different rate schedules within each class. This discussion only addresses residential customers, so if you're concerned about your commercial or industrial use, please seek help in determining your savings and options you may have to save money and the planet.

- Some utilities may charge different rates for electricity consumption during different times of the day. It generally costs more for utilities to generate power during certain periods of high demand or consumption , called peaks.
- Some utilities may also charge a base rate for a certain level of consumption and higher rates for increasing blocks of consumption.
- Often a utility adds miscellaneous service charges, a base charge, and/or taxes per billing period that could be averaged per kWh consumed, if these are not already factored into the rate.

What you, as a residential user, need to do is just look at your bill. It has to show your costs for all factors. If you can't figure it out, call the utility company, and have them tell you what it is. They're required to do this, so don't let them just say to look at your bill again.

LED LIGHTING:

A light-emitting diode (LED) is a semiconductor (56) device that emits visible light when an electric current (57) passes through it.

Both CFL (above) and LED lighting conserve a lot more electricity than the other common options. The main difference is: LEDs are more expensive, but can be dimmed and do not contain mercury. Cost savings isn't worth the significant environmental problems caused by mercury, so I hope you'll consider LED as your light source whenever possible.

Other advantages of LED:

Unlike fluorescent lamps, the operating life of a light emitting diode (LED) is unaffected by turning it on and off. This characteristic gives LEDs several distinct advantages when it comes to operations. For example, LEDs have a huge advantage when used in conjunction with occupancy sensors or daylight sensors (58) that rely on on-off operation. Also, in contrast to traditional technologies, LEDs turn on at full brightness almost instantly, with no delay. LEDs are also largely unaffected by vibration because they don't have filaments or glass enclosures.

Calculating Energy Savings:

To calculate the exact value of energy savings by turning a light bulb off:

1. First, determine how much energy the bulb consumes when on. Every bulb has a watt rating printed on it. For example, if the rating is 40 watts, the bulb will consume 0.04 kWh in one hour. Or, if it's off for one hour, you'll save 0.04 kWh. (Note that many fluorescent fixtures have two or more bulbs. Also, one switch may control several fixtures—an "array." Add the savings for each fixture to determine the total energy savings by switch array.)

2. Then you need to find out what you're paying for electricity per kWh (in general and during peak periods). You'll need to look over your electricity bills (as noted above in the CFL section) and see what the utility charges per kWh. Multiply the rate per kWh by the amount of electricity saved, and this will give you the value of the savings.
3. Continuing with the example above, let's say your electric rate is 10 cents per kWh. The value of the energy savings would then be 0.4 cents ($ 0.004) per hour for turning off that bulb. The value of the savings will increase the higher the watt rating of the bulb, the greater the number of bulbs controlled by a single switch, and the higher the rate per kWh.
4. To give you a hint of the financial impact, let's say that light is usually left on all night. For simplicity, we'll stick with just the single bulb. If you factor $.004 times eight hours in a night, that's $.032. That might not seem like a lot, but adds up to almost twelve dollars of waste every year. For just one light bulb.

The most cost-effective length of time that a light (or set of lights) can be turned off before the value of the savings exceeds the cost of having to replace bulbs (due to shortened operating life) will depend on the type and model of bulb and ballast. The cost of replacing a bulb (or ballast) depends on the cost of the bulb and the cost of labor to do it.

Lighting manufacturers should be able to supply information on the duty cycle of their products. In general, the more energy-efficient a light bulb is, the longer you can leave it on before it's cost effective to turn it off.

Automatic lighting controls:

In addition to turning off your lights manually, you may want to consider using sensors, timers, and other automatic lighting controls. (59) Even though purchase and installation can be a little expensive, if you're not handy at this kind of thing, automatic controls can take the struggle out of managing light efficiency in a house full of people. Many systems now even provide a computer control that can be managed via your phone. I personally have a system like that, which has really helped control certain lights that seemed to be on at night a lot. Now, I can check them from my phone and turn them off prior to going to bed without having to go through the whole house spot-checking. Okay, I'll admit that I've occasionally put someone in the dark for a moment (until the inevitable shout), but that minor inconvenience is small price to pay.

The calculation of your actual possible savings starts with determining what type of bulbs you have and their respective watts. Next, determine replacement costs for bulbs and any installation that may be required. I know this is a little work but it'll be worth it. With the change-out information, the best way to figure out your savings is to go to the web site below that provides a calculator to plug in all your information.

Light bulb cost calculator available from:
https://www.inchcalculator.com/lighting-energy-cost-calculator/

WHY IT MATTERS:

Controlling on and off times is important but, when possible, the real savings in home lighting is in changing the bulbs you use. Power usage beyond the lighting is focused on the appliances you use in your everyday living.

- The DOE estimates that an average house contains around 45 light bulbs. In my experience, this count is low for a typical home of 1800 square feet or more, when you count the actual number of bulbs rather than just the number of fixtures.

- Lighting thus makes up a significant part of your average utility bill. (60) In fact, Americans spend up to 25% of their electrical utility bill on lighting.

With a little up-front cost and effort, by switching to LED and turning off lights that aren't needed, you can really make a dent in your electric bill and long-term cost of bulbs, not to mention reduce strain on your local power grid. Add a few extra automatic controls, and you can keep saving with even less work and stress.

Give it a try and see how much you can save, then tell your friends and family, so they can do the same. The money and power you'll save — not to mention, no more piles of old-school light bulbs cluttering our landfills — are well worth the effort.

13
Appliances

This activity is a little more challenging because it may take a significant investment to make the changes from current appliances but the good news is that they will pay you back and provide even more savings long term.

On average, home appliances—including clothes washers, dryers, dishwashers, refrigerators, freezers, air purifiers and humidifiers—will account for 20% of your home's total electric bill. ENERGY STAR appliances, [61] which are certified by the U.S. Department of Energy, can reduce that share. The average home appliance lasts for 10 to 20 years, and an ENERGY STAR certified appliance will use anywhere from 10 to 50% less energy each year than a non-energy efficient equivalent.

WHAT YOU CAN DO:

Choose energy-efficient appliances with the Energy Star Label when you replace old ones:

While most energy-efficient appliances cost more than their less efficient counterparts, the savings to your monthly utility bills will add up. ENERGY STAR appliances can significantly reduce your electric bills. For more details, go to:
https://www.energysage.com/energy-efficiency/costs-benefits/energy-star-rebates/

How much can you save?

Clothes Dryers:

Dryers are the most energy-hungry appliances in the average American home. According to the Natural Resources Defense Council (NRDC), a typical dryer can consume as much energy per year as a new energy-efficient refrigerator, washing machine, and dishwasher combined—and if you have an older model, that number could be even higher. ENERGY STAR certified dryers use 20% less electricity than a conventional model, which will save you $210 in electric bills over your energy-efficient dryer's lifetime.

Clothes Washers:

If you pair your energy-efficient dryer with an energy-efficient washing machine, you'll see additional savings on both your electric bill and your water bill. An ENERGY STAR certified clothes washer uses 40 to 50% less energy and about 55% less water than standard washers.

When you replace your conventional washing machine with an energy-efficient model, you can expect to save up to $50 per year on utility and water bills. Savings from changing out older models can add up faster. Replacing a pre-1994 washer with a new Energy Saver model can save an average family of four about $110 per year on their utility bills. Energy-efficient front-loading clothes washers also require less laundry detergent than top-loading washers, so you'll save even more money week to week.

Refrigerators:

There have been significant advances in refrigerator technology over the past fifteen years, which means that old refrigerators are one of the biggest energy hogs in many U.S. homes. The ENERGY STAR certified refrigerators available on the market today are nearly 10% more energy efficient than models that meet the federal minimum energy standard. If you have an older refrigerator, you can save even more on your energy bills with an energy-efficient refrigerator. ENERGY STAR qualified refrigerators use up to 40% less energy than the conventional models sold in 2001.

Dishwashers:

Dishwashers may not use as much power as a constantly running refrigerator or high-heat clothes dryer, but the electricity and water that's needed to run a dishwasher cycle adds up. ENERGY STAR certified dishwashers are 12% more efficient than non-certified models currently for sale, and installing an energy-efficient dishwasher will save you around $25 a year.

WHY IT MATTERS:

By replacing the appliances in your home or business with ENERGY STAR certified models, you're making an investment that'll reduce your energy bill for years to come, which is especially important when you recognize that electricity rates are increasing every year,[62] That being said, some appliances use more energy than others. The amount you save is also dependent on the age of your current appliances and the electricity rates that you pay. Old refrigerators, for example, use up to 50% more electricity than newer models.

Okay, now you have some data to work with, options to decide what works for you, and the savings you can achieve while saving the environment. So get out there and start the changes and figuring out what you can do with the real money you'll save in this one area, alone. And, don't stop there. Let your friends and family know the money and resources they can save by switching to ENERGY STAR appliances, too.

14
Options For Less Driving

This activity takes some commitment and a real change in lifestyle but again can make a real contribution to the wellness of the environment and will save money. The possible options include carpools, public transit, better scheduling of car usage for personal errands, and telecommuting. As to telecommuting the other factor often missed is the true value personally of telecommuting when possible to most people with a family at home or not. Studies actually show telecommuting can in fact reduce your personal stress if managed effectively. Starting off with one fact showing how impactful a change could be is that if you can leave your car at home or limit your weekly outings for any reason and stay off the road just two days a week, you'll reduce greenhouse gas emissions by an average of 1,590 pounds (721 kilograms) per year, according to the EPA. That's a lot for just one person.

The first part of this activity is the considering the carpool option. Carpooling requires a change of driving habits that may be tougher for some folks to do than others. Between the demands of work and raising kids, not to mention desired flexibility in your own transportation, it can be a real challenge to organize and maintain a functioning carpool. Even combining errands or arranging to telecommute can be hard to manage. But these shifts in car use can save gas money and help reduce your energy footprint, along with actually getting to know your neighbors or co-workers. While we all might not look at spending more time with our office-mates as a positive, my experience with rare exceptions has been rewarding on many levels.

WHAT YOU CAN DO:

Carpool and ride-share:

- Check with coworkers, to see who commutes from your local area, and arrange a schedule to share rides whenever possible.
- Include as many people as comfortably possible, so everyone gets the benefit of slack time, keeps their gas use for work commutes low, and can easily absorb individual scheduling conflicts.
- If a work-mates carpool isn't possible, maybe you know neighbors who work in the same general vicinity, and can talk with them about arranging a ride share.
- Check with fellow parents at schools or day-care to organize a carpool for your younger generation, as well.
- Even chore day can have a carpool. Do you have neighborhood friends you can run errands with and share shopping trips?

Pros and Cons:

Pros:

- Carpooling can save a lot of money on gas, plus wear and tear on your vehicle. With gas prices hovering around all-time highs, it's always nice to save a few gallons a week by driving to work with others.
- By spending more time with folks from work, you'll have time to talk and get to know them better. Not only can you possibly discover things in common you hadn't expected, the new camaraderie can make time at work more pleasant, spent with people who are honest-to-goodness friends.
- Not only will you save tangibles, like gas money, but you'll also reduce stress. Sharing driving responsibilities gives you the gift of not dealing with

traffic, so you can relax and enjoy your morning coffee with no pressure.
- And, if you're the social type, you'll have fresh conversations about news, life, and the latest hot shows and movies to look forward to every day.

Cons:

- Carpooling is a commitment, and requires spending extra time with people from work, so be careful who you invite. If you unknowingly pick someone who enjoys spreading office gossip, it can cause trouble on several levels.
- A carpool is only a carpool if everyone takes a turn. If you choose a carpool mate who doesn't pull their weight, and the others must drive more than their share, it can cause bad feelings that could carry into your workplace.
- If someone calls in sick or has a last-minute emergency, especially that day's driver, the rest of the carpool needs to regroup on the fly. But you can offset much of the potential hassle with a little organization. When you set up your carpool, make sure to form a few contingency plans, in case something goes awry. Even if the backup plan is just to call a Lyft or Uber, take the time to coordinate beforehand, so you're ready for any possible snags in your schedule.

Keep your car at home:

Sometimes, the best solution is simply not going by car. Sure, the trip will take a little more time, but getting out on your own can be great fun and gives you a chance to see more of the neighborhood you live in. Plus, you'll burn some calories and reduce greenhouse gases, too.

- Walking or riding your bike to work, school and anywhere you can is great both physically and energy saving.
- If the locations and services you need are a bit further, try using mass transit.

Combine your errands:

When the car is a must, try combining your errands and map them out to make the most efficient use of your time and gas. For example, you could drop the kids off at their extracurriculars before hitting the hardware store, mall and supermarket, and then pick them back up on the way home. I know it can take a while to organize the troops if other family members need to come along, and with kids' practices and weekend plans, it may not always be practical. But, you'll never know how well it can work until you try.

Work from home:

In some jobs, it may even be possible to work at home one or more days per week. If you have the kind of job that you can do remotely, check whether your company would allow you to telecommute. More companies are discovering that letting some of their staff work from home is not only better for their bottom line, but also great for productivity and morale.

If your company doesn't already offer telecommuting, it may be tough to convince them, at first. Unfortunately, some employers still think you'll be avoiding work somehow. A great discussion about how you can start persuading your boss written by Bill Murphy Jr., a contributing editor for Inc. Com., can be found at:

https://www.inc.com/bill-murphy-jr/7-steps-to-convince-your-boss-to-let-you-work-from-home.html

WHY IT MATTERS:

There are currently around 700 million cars on the road producing 900 million tons of carbon dioxide (63) each year. This equals approximately 15% of our total output. Sadly, one half of these trips in the U.S. are within 3 miles, and could easily be walked in less than an hour. If the number of cars keeps increasing at its present rate, there'll be over one billion on the road by 2025.

Remember that figure at the top of this section? That, by leaving your car at home just two days a week, you'll reduce greenhouse gas emissions by an average of 1,590 pounds (721 kilograms) per year. Now, instead of factoring only yourself, imagine how much good you're doing for the environment when you and three work mates team up to share rides.

So, you have a few things to consider and, as always, start with determining what your current costs are for the last six months for transportation. Then keep track for the next six months, and you should be able to see a tangible change in expenses and calculate your emission reduction (if you are into it like I am). The intangible stuff like getting to know people and your neighborhood — along with maybe even feeling healthier if you start riding a bike or walking more — is tough to quantify but, trust me, they're real positives.

You'll never know how well something will work until you give it a shot. Once you start, and the benefits kick in, you'll want to share the wealth of savings and fitness.

15
Used Motor Oil

This activity is close to my heart, because used oil not managed correctly can cause significant harm to the environment and really cause significant harm to overall sustainability.

One gallon of motor oil (64) dumped on the ground or in the trash can contaminate up to a million gallons of fresh water, which is a year's supply of fresh water for fifty people. I do a lot of compliance support with people who manage used oil for a living and sincerely believe they perform an invaluable service to their community and this planet. Proper disposal is critical, as the U.S. generates a total of 1.3 billion gallons of waste oil each year, of which only 800 million gallons are recycled and 500 million are disposed of improperly. And, looking at the problem from a conservation perspective, that's a tremendous amount of available oil, if we could just get it to the facilities.

An editorial note: most of the information in this activity discussion comes directly from state and federal agency sources, and I encourage you to go to any state site or the EPA site and do some more reading on the issue.

WHAT YOU CAN DO:

Consider going to an oil change pro:
If you're one of the many Do-It-Yourselfers (DIY) who change their own motor oil, you need to know how to properly manage the used oil. At the start, I must suggest that if you can afford it, I'd like you to really consider going to your local oil change place instead. I don't feel the same about all car repairs, but most professional oil changes are rather cheap, compared to other costs in vehicle ownership, and all of them must always follow the proper requirements.

Businesses that offer oil changes need to comply with stringent oil management regulations set by the government or face strict penalties. They already have foolproof systems in place to ensure compliance for safe and environmentally sound handling.

Considerations to DIY oil changes:

If you still want to do it yourself, when handling used oil, be sure to take these key points into consideration:

- Used motor oil is insoluble, persistent, and can contain toxic chemicals and heavy metals like lead and other real nasty stuff that can cause more than a little harm to you and the environment.
- It's slow to degrade.
- It sticks to everything, from beach sand to bird feathers as well as you, your clothes and towels, and anything else it touches.
- On average, about four million people reuse motor oil as a lubricant for other equipment or take it to a recycling facility. If you plan to recycle your used oil, be very careful not to spill any when you collect it and place it in a leak-proof can or container. (Specific handling suggestions are provided in steps below.)
- Check with local automobile maintenance facilities, waste collectors, and government waste officials to see when and where you can drop off your used oil for recycling.
    a) Don't forget to drain and recycle used oil filters as well—usually you can drop off the filters at the same collection centers where you deposit used oil.
    b) To find a used oil recycler, if you don't know of one, go to https://search.earth911.com for the information based on your zip code. Plus, the site also has helpful information on other recycling.

Used oil management step-by-step:

If you change your own oil on your car, truck, motorcycle, boat, recreational vehicle or lawnmower, be certain to work carefully and dispose of the used motor oil and filters properly. Follow these steps for a clean oil change that prevents pollution and conserves energy for a safer and healthier tomorrow.

The Minnesota Pollution Control Agency has great help pages you can go to for additional guidance. Information for the guidelines used here can be found at: https://www.pca.state.mn.us/living-green/changing-your-oil-earth-friendly-guide-do-it-yourselfers

An important note: safety first:

What's not mentioned in the original state advice is: put on some safety glasses as well as safety gloves that won't allow the oil to get onto your skin. Both safety glasses and gloves can be bought pretty cheaply at your local hardware store.

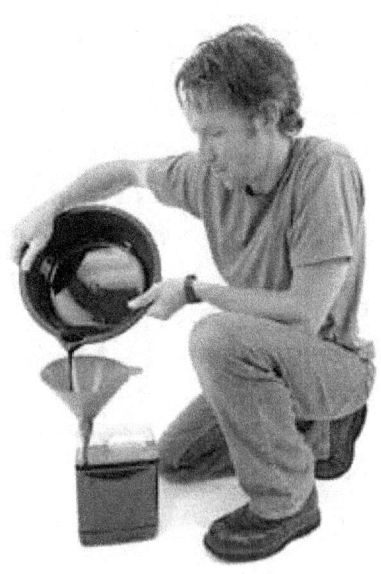

Step 1:

Drain the oil:

- Drain the oil into a pan that can hold twice the volume of oil in the engine's crankcase.
- Drain the oil when the engine is warm to ensure that any sludge flows out smoothly. Caution: the oil will be hot! Keep draining until the oil has slowed to an intermittent drip.
- Replace the drain plug and carefully move the oil pan to a location where you can safely pour the oil into a container.
- Wipe up any drips with a paper towel.

Step 2:

Carefully transfer the dirty oil:

Using a funnel, pour the oil into a clean leak-proof container with a tight-fitting lid—a rigid plastic container such as a plastic milk jug works well. Be careful not to overfill the container.

- Don't ever put used oil in containers that held chemicals like bleach, pesticides, paint, or antifreeze—they can contain residues that contaminate the oil. Avoid paint cans and other metal containers.
- Seal the bottle and label it as "Used Oil."
- Power steering, transmission, and brake fluids can also be brought to used oil collection sites in separate containers (not mixed with used oil).
- Never mix solvents, gasoline, or antifreeze with your used oil. Once contaminated with these products, it's difficult or impossible to recycle used motor oil.

a) If your oil does become contaminated, label the container and take it to your local Household Hazardous Waste collection site for disposal. Don't take it to an oil recycling site where it could contaminate the tank, making the contents impossible to recycle and expensive to dispose of.

Step 3:

Drain the oil filter:

- Put the used filter hole-side-down to allow the oil to drain for recycling.
- Allow the filter to drain overnight or at least 12 hours to remove all the oil.

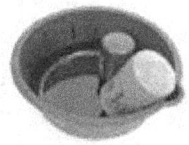

- Place the old filter in a leak-proof container—a coffee can with lid or resealable plastic bag works well. The oil filter may contain 2 to 8 ounces of motor oil, even when drained.
- Replace the old oil filter with a new one.
  a) To support recycling programs, purchase your oil filters from businesses that also accept used oil filters for recycling.

Step 4:

Take your used oil and oil filters in for recycling:

- Find a designated collection site in your area. (Earth911 – noted above – is a good source for locating a recycle facility.)
  a) If you're in Minnesota, all places that sell motor oil must post a listing of used oil collection site locations or a toll-free phone number with that information. (This is unfortunately not yet required in all states)
- Many oil change shops and service stations with repair facilities will accept your used oil, sometimes for a fee.
- Communities and counties often provide collection tanks for used oil.
- Don't leave used oil or oil filters at a collection facility if it's not open for business unless there are specific instructions at the site that allow you to do so. This is considered illegal dumping.

WHY IT MATTERS:

Some facts about used oil and filters:

- Nearly 40% of the pollution in America's waterways is from used motor oil.
  a) Oil finds its way through any kind of water flow into streams, rivers, bayous and canals, and a lot of it ends up in our oceans and lakes.
  b) Oil pollution isn't just from dumping. Runoff by vehicles not well maintained also contributes to the problem.

- Used oil filters pose similar waste concerns.

a) If properly drained, they can be safely recycled or disposed.
b) Each year, the United States generates 425 million used automotive oil filters containing 160,000 tons of iron units and 18 million gallons of oil.
c) Used oil filters can contain more than 45% used motor oil in weight when removed from the vehicle.

Benefits of Reusing and Recycling Used Oil and Filters:

- Recycled used motor oil can be re-refined into new oil, processed into fuel oils, and used as raw materials for the petroleum industry.
    a) Re-refined oil must meet the same stringent refining, compounding, and performance standards as virgin oil for use in automotive, heavy-duty diesel, and other internal combustion engines, and hydraulic fluids and gear oils.
- Extensive laboratory testing and field studies conclude that re-refined oil is equivalent to virgin oil—it passes all prescribed tests and, in some situations, even outperforms virgin oil.
- Recycling used oil keeps it from polluting soil and water.
- Motor oil doesn't wear out—it just gets dirty—so recycling it saves a valuable resource.
- Less energy is required to produce a gallon of re-refined base stock than a base stock from crude oil.
- One gallon of used motor oil provides the same 2.5 quarts of lubricating oil as 42 gallons of crude oil.
- Used oil filters also contain reusable scrap metal, which steel producers can reuse as scrap feed.
- The same consumers and businesses that use regular oil also can use re-refined oil, since re-refining simply re-processes used oil into new, high-quality lubricating

oil. In some cases, fleet maintenance facilities that use large volumes of oil arrange to reuse the same oil that they send to be re-refined—a true closed recycling loop.
- If we re-refined the 1.3 billion gallons of waste oil generated in the United States each year, we would save 1.3 million barrels of oil per day, or half the daily output of the Alaska pipeline.

Now that you have the information about why it's important to handle your used oil the right way, you can decide whether doing it on your own or taking your auto to the local oil change place is the way you want to go. The smaller oil-containing equipment you may have at home will still need your personal attention, but now you have some suggestions and agency direction on how to do it as well. Just be safe.

Please, bear in mind: this is not really an appropriate activity for small children, though it's a great opportunity for teaching responsible maintenance. So, if they're interested, allow them watch from a safe distance. If older kids have an interest or you just need the help, they should be fine—with your supervision, of course.

When kids watch or help, make sure they know what you're doing and why. Tell them what you're going to do with the oil to keep it from contaminating our environment and hurting anyone. Maybe even take them along to the recycler to show them the whole process.

The time and care you spend in used oil management today can help conserve oil reserves, plus give us and our environment a lot cleaner and healthier tomorrows.

"Because we don't think about future generations, they will never forget us."
— HENRIK TIKKANEN

16
Plant a Tree

This activity can really be a great activity for the whole family to enjoy is planting a tree, which can be great fun and provide rewards for everyone for a lifetime. Even if you and your family don't stay in your present house forever, the next owner will appreciate it and, in most cases, trees actually add value to your home. So, let's see how it's done.

WHAT YOU CAN DO:

Plan before planting:
  Planting a tree is more than just popping a plant into the ground. Before you dig in, there are a few decisions to make and details to consider:

Type and size of tree:

- Size is a consideration when planting trees, because they vary in height and spread, which will determine the space you should leave between them and your home or outbuildings to accommodate for its reach when it's full size.
  a) Large trees, up to 70 feet or more, need at least 20 feet.
  b) Medium-sized trees up to 70 feet tall, need 15 feet.
  c) Small trees 30 feet tall or less, need 8 to 10 feet.
- Root systems are also a factor. If you want to plant a tree closer than 20 feet from your house, it's best to plant a smaller tree that has well-behaved roots.

- In selecting a tree and location to plant, remember to consider proximity to the house from a safety perspective, too. Especially in a windy region like the Gulf of Mexico area, where hurricanes are a common occurrence, windblown trees can cause real harm to the house and, worse, your family.
- If you want plant a fruit tree, I encourage you to think of the eventual reality before deciding a type and location.
  a) Are you prepared to handle a tree's worth of fruit that must be picked or cleared from the ground?
  b) Will the tree produce messy fruits that could stain sidewalks and driveways?
  c) Do the fruits have pits that could remain hidden and hurt bare feet if left on the ground in a grassy area?
- Hickory trees, sweet gums, oaks, and similar trees drop hard fruit (nuts, pecans, acorns, seed pods, etc.) that can be a hassle to clean up.

Where to plant:

- Choose your site carefully, based on the type and expected size of the tree you're going to plant. From personal experience, moving the darn thing can be a real pain.
- Make sure you consider tree root issues. Root systems can damage sidewalks and foundations — which is the reason I needed to move the tree mentioned above.
- Is there enough sunlight for proper growth where you'd like to plant?
- As I mentioned above, always remember to think of safety when planting a tree. If you live in a windy area or region prone to hurricanes, windblown trees can damage your house or harm your family.

Before you dig:

It is the law in all 50 States that YOU MUST CALL 811 BEFORE DIGGING to protect yourself and others from unintentionally hitting underground utility lines. There are millions of miles of buried utilities beneath the surface of the earth that are vital to everyday living like water, electricity and natural gas.

- 811 is a free service managed by Underground Service Alert and available to everyone. Whether you are planting a tree or a garden, or digging holes for fence posts, call 811 at least two business days before you plan to start your project.
- Go to the 811 CALL BEFORE YOU DIG web site at https://call811.com/Before-You-Dig for more information.

Plant your tree:

- Read information, view a video on planting, or consult a professional to guide you and ensure you know the correct depth for planting your tree. Following are some websites and videos for reference (plus a lot more information, if needed):
  a) The Arbor Day Foundation is a 501c3 nonprofit conservation and education organization. A million members, donors, and partners support their programs to make our world greener and healthier. Info available at https://www.arborday.org/celebrate/tree-planting.cfm
  b) Lawn Care Academy the "Complete Tree Planting Instructions—The No Regrets Tree Guide" available from https://www.lawn-care-academy.com/tree-planting-instructions.html

c) STIHL USA tree planting video available at
   https://www.youtube.com/watch?v=0lY2qrtPZR0
d) Cornell Plantations "Trees: How To Properly Plant a Tree Cornell Botanic Gardens" available from
   https://www.youtube.com/watch?v=C3kazAcJZlU
e) Frank Otte Nursery and Garden Center info available at
   https://www.youtube.com/watch?v=SlUQK8GPp5w

- Make sure everyone dresses for gardening work and wears gloves while doing the planting.
- The diameter of the hole is important. Again, view a video, as there is some conflict in the old adage of digging the hole twice the diameter of the root ball.
- Soil quality is important, so consider mixing some compost with the backfill to improve soil quality, if needed.
- Tree planting instructions are dependent upon the types and quantities of trees you want to plant. I've only provided a limited list of possible considerations and things to be aware of, but the list of websites and videos above can give you more detailed guidance on planting a tree.
- Watering Newly Planted Trees: In general, water a tree once a day for the first two weeks, and after that once a week for a year, as long as the tree is not dormant (without leaves). When trees are newly planted, their watering requirements are high.

WHY IT MATTERS:

Reasons easier to explain to younger kids:

- Trees give off oxygen that we need to breathe. One large tree can supply a day's supply of oxygen for four people.
- Trees create a place to live (ecosystem) plus food and protection for many birds and mammals.
- Trees can provide shade to play under when they grow tall.
- Fruit trees can provide fruit for us to eat.

Reasons suitable for older kids:

- Trees reduce the amount of storm water runoff, which reduces erosion and pollution in our waterways and may reduce the effects of flooding.
- Trees absorb carbon dioxide and potentially harmful gasses, such as sulfur dioxide, carbon monoxide, from the air and release oxygen.
- Trees improve air quality, climate amelioration (by helping offset some negative aspects and causes of climate change), water conservation, and soil preservation.
- During the process of photosynthesis, trees take in carbon dioxide and produce the oxygen we breathe, as noted above.

Reasons you'll like:

- Property value: The proximity of healthy, beautiful trees increases property value. Good tree cover increased property prices by about 7% in residential areas and 18% for building lots.
- Noise reduction: North Carolina State University showed that a well-placed tree can block noise by as much as 40%.

Now, review some of the videos and other information provided and get out there and plant a tree. It's a simple process for you and the kid(s) to do, accomplishes so much for the environment of both your immediate habitat and the larger world, plus is a fun way to teach a living example of environmental stewardship. And, as a bonus, your whole family will have a beautiful tree (and maybe fruit, too) to enjoy for years to come.

SECTION THREE: ADVANCED ACTION

"Because we don't think about future generations, they will never forget us."
~ HENRIK TIKKANEN

So far, you've gotten your feet wet with some relative minor changes of habit, and then taken your sustainability action a step further to move forward into more lifestyle and long-term activities. Now it's time to get to some advanced action.

Some activities might involve a simple matter of more mindfulness in purchasing by individuals and manufacturers that impact consumption decisions, while others require a little more focused action for you to be informed more on where you buy things and who supplies them, and maybe digging a little deeper into the budget short-term to ensure long-term gains for you and Mother Earth. Regardless, all the activities here have the potential to turn a little extra effort into a huge impact.

It's a fact of life that, once our society has established a convenience, it's next to impossible to let it go. So, how can we have our modern lives, transportation, machines, and time-saving appliances while preserving and restoring critical elements of our environment? What products or processes can we rethink or improve to support our smaller efforts?

Here, we'll focus more on the bigger picture. What are you doing to, not just stem the tide of waste and pollution in our current lifestyles, but plan for sustainability? How can your actions today, rather than mitigate harm, lay the groundwork for improvement?

The changes you make today can be the foundation of a much better tomorrow.

17
Buy Local

As a nation, we've developed a harmful habit of chain retailers and big box super stores, which may seem more economical in the short-term, but more often have a devastating effect on local economies and the environment. While the mega-retailers rely on imported goods, bloated supply chains and fleets of trucks to operate, locally owned businesses make more purchases from other local businesses, service providers and farms. Their smaller business model, rather than draining the economy and leaving pollution behind, keeps profits in the community and requires less transportation. Their leaner model also results in reduced traffic congestion which results in lowered fuel emissions and air pollution as well as less packaging and waste straining our already-overburdened landfills.

The impact of buying local is great for our wallets, too. Small local businesses are the largest employers nationally and create two out of every three new jobs and employ more than 52% of the nation's employees. And, because they also buy local, more money circulates in the local economy, which helps support public infrastructure like libraries and schools, and government services.

Buying local helps responsible businesses grow, as well as the local tax base, and develops a local economy that's both economically and environmentally sustainable.

WHAT YOU CAN DO:

There are a lot of ways to integrate buying local into your routine. Instead of the national chains and mega-retailers, try locally owned and supplied providers:

- Supermarkets, farmers' and ethnic markets, and niche food shops

- Restaurants, diners, and fast-food shops
- Specialty retail: like tools, building and hardware, housewares and supplies, clothing, books, electronics, appliances
- Service providers: like printers, cleaning crews, garden and landscaping, pest control, pharmacists, and movers
- Entertainment and recreation activities

You might pay a little more for your goods and services, in the moment. But the boost to your local economy and environment will reap long-term rewards for the whole community.

WHY IT MATTERS:

Facts and misconceptions:
Buying local is not only an economic benefit to the local businesses and people in your community. It has significant positive environmental impacts, especially associated with sustainability, backed up by quite a lot of data and writings. In the various category sections below, I've summarized information from an article provided by Michigan State University Center for Community and Economic Development September 2010 principal author Nandi Robinson and contributor Rex L. LaMore, Ph.D. that discusses the primary issues, why this is so important for you to do, along with my comments and amplification on a few issues provided within parenthesis (like this).

The primary question — or misconception — is best answered by Michael H. Shuman, author of the book Going Local. "Going local doesn't mean walling off the outside world. *(Buying locally doesn't create an isolation mentality, nor does a macro position of buying American whenever possible.)* It means nurturing locally owned businesses that use local resources sustainably, employ local workers at decent wages and serve primarily local consumers. It means becoming more self-sufficient and less dependent on imports. Control moves from the boardrooms of distant corporations and back into the community where it belongs." (Shuman 2000)

Sustainable local economy:

Small local businesses are the largest employers nationally and create two out of every three new jobs. The Small Business Act defines a small business as "one that is independently owned and operated and which is not dominant in its field of operation." Small businesses employ more than 52% of the nation's employees. This means that, overall, more Americans work for a company with fewer than 100 employees than for a large retailer with more than 500 employees. Small businesses play a vital role in job creation, adding more than 5.1 million new jobs to our economy since 2003.

Buying locally means that employment levels are more likely to be stable, and may even create more opportunities for local residents to work in the community. *(This one fact represents perhaps the most important aspect of your efforts in community support, using your own economic power to make a change that can impact you and your neighbors directly in a short time line.)* When dollars are spent locally, they can in turn be re-spent locally, raising the overall level of economic activity, paying more salaries, and building the local tax base. This re-circulating of money leads to an increase of economic activity, with the degree of expanse entirely dependent on the percentage of money spent locally.

The Local Premium:

The Local Premium represents the quantifiable advantage to the city provided by locally owned businesses, relative to chain businesses. It's the added economic benefit of local businesses to a local economy. According to the Andersonville

Study, local businesses generate a substantial local premium, or added economic benefit over chain retailers. *(Yes, I realize chain retailers, big box and super stores are often more economical in the short term. But it's an emotionally fraught subject, as their entry into communities can devastate local business and change main streets once alive with unique charm into homogenous every-towns. If we actually support our historical main street businesses instead, positive impacts including economic value can be achieved in a relatively short time—but we all have to participate.)*

Thriving local business means more money circulating in the local economy, which leads to more public infrastructure like libraries and schools, plus raises more money in taxable transactions to fund local government services. The Local Works West Michigan Economic Analysis describes four ways in which a firm keeps money local: wages and benefits paid to local residents, profits earned by local owners, the purchases of local goods and services for resale and internal use, and contributions to local nonprofits. Consistently, locally owned businesses exceed their chain competitors in all four components. (This is an important fact, despite what many of the super stores promoters would want you to believe.)

Case study: Grand Rapids

The MSU article displays the recirculation of money into the Grand Rapids economy by a locally owned business and its non-locally owned competitor *(and the data is typical for many small community economics)*. Significantly more money re-circulates locally when purchases are made at the locally owned business. This recirculation is attributed, in part, to locally owned businesses purchasing more often from other local businesses, service providers and farms. Purchasing locally helps other businesses grow, as well as the local tax base.

According to the Local Works analysis of the West Michigan economy, for every $100 in consumer spending with a locally owned business, $73 remains in the Grand Rapids Economy. The remaining $73 is then dispersed locally in the form of wages, charitable donations, taxes which fund city services, and purchases of goods and services from other local businesses.

On the converse, for every $100 spent at a non-locally owned business, only $43 remains in the Grand Rapids economy. When economic stimulus comes from outside of an economy *(e.g., tourism, federal funding, and industrial exports)* the full effect of those dollars depends on how much of that money remains in the local area.

Sense of community:

Spending money with local retailers also helps keep the local community alive. The places where we eat, shop, and have fun all make a community feel like home. "One-of-a-kind" businesses are a fundamental part of the distinctive character and of a community. Conversely, a community where large chains exceed the number of independently run small businesses becomes less personal and more homogenized, with limited product diversity.

Equally important to economic benefit is how thriving independent businesses define a community's self-image and create a sense of pride for the people who live there. National chain retailers, on the other hand, reduce a community's uniqueness and originality. *(Despite a move to make all of us in the same mold, we have and should have regional perspectives and needs as individual local communities. We can simultaneously share a common set of beliefs as Americans while still celebrating our uniqueness as a community.)* More and more communities are choosing to retake control of their own economic character by supporting unique one-of- a-kind local businesses.

Consumer value and choice:

A marketplace of hundreds of small businesses is the best way to ensure innovation and low prices over the long term. *(In English, this means the marketplace should control the prices, not some big mega-store setting the prices because they've forced out all the smaller local stores.)* A multitude of small businesses, each selecting products based on their own interests and the need of their local customers, not a national sales plan, guarantees a much broader range of product choices. Plus, a thriving local economy attracts even more talent and innovation. A growing body of research shows that, in an increasingly homogenized world, entrepreneurs and skilled workers are more likely to invest and settle in communities that preserve their one-of-a-kind businesses and distinctive character with multiple consumer choices.

Environmental impact:

Buying local also positively impacts the environment by lessening the amount of materials and products bought from national retail chains, which helps reduce your ecological footprint. Locally owned businesses make more local purchases, requiring less transportation, and generally operate from within city centers, as opposed to developing on the outskirts of a city. More central commercial districts results in fewer vehicle miles traveled and less sprawl. Less transportation also means less traffic congestion, which has the potential to reduce the amount of fuel emissions that contribute to air pollution.

Locally sourced materials and products have added environmental benefits. While eliminating unnecessary transportation and delivery, they produce less waste by reducing the amount of packaging being used. Less packaging waste also means less demand on landfill sites. According to the National Resource Defense Council, buying local will help reduce pollution, improve air quality and improve our health.

Chain Retailers: The False Promise of Economic Growth:

While national businesses have a role to play in every economy, purchases from national businesses typically cause money to leak out of the local economy. National chains send money outside of the community to the areas where they are headquartered. Plus, large chain retailers often draw revenues from neighboring communities, so that even towns adjacent to locations with new chain retailers see sizable losses in both sales tax revenues and employment, according to The Santa Fe Independent Business Report. With large national businesses growing in both numbers and employment totals at rates much faster than smaller businesses, their arrival also wreaks devastating effects the small business sector through increased competition for labor, higher rents, and usually a decrease in small business sales.

Sustainable economic growth:

The premise that locally owned and operated businesses generate greater local economic activity than their chain counterparts has become widely understood and accepted. In communities across the nation and abroad, public policy has adapted to this reality through a variety of planning and zoning tools. Local economic growth also attracts new talent and professionals who may, in turn, create businesses of their own, enhancing a local economy.

According to the Small Business Association of Michigan (SBAM), Michigan must begin to pursue a culture of entrepreneurship to stimulate more individuals to create their own growth-oriented firms and to nurture the existing firms. This nurturing has been coined "economic gardening" by the SBAM and is a new approach to economic development which focuses on strengthening small firms positioned for growth rather than trying to recruit or retain companies that could locate elsewhere, like national retailers.

The most valuable asset to the pursuit of an entrepreneurial culture is college graduates. These young professionals are attracted to urban communities characterized by complex public transportation systems, residential and retail developments, and a variety of amenities like restaurants, bars, and galleries all within a densely populated community. The economic growth resulting from these locally owned businesses helps to expand community areas.

Health benefits:

Along with the information from Michigan provided above, I want to address one more issue: your health. According to the Cambridge Journal of Regions, Economy, and Society, [77] researchers who studied 3,060 counties and parishes in the U.S. found that counties with a higher volume of local businesses actually had a lower rate of mortality, obesity, and diabetes. So, if you can't rationalize why eating and buying foods locally makes economic and environmental sense, know this fact: eating foods that are unadulterated with pesticides like organic fruits and vegetables, or free from hormone disrupting compounds like grass-fed meats, pasture-raised eggs and dairy, contribute to a huge upgrade in the Standard American Diet.

Local is sustainable:

You've been given the info, so now you know the impact you can have by changing your buying habits in your community. Yes, in the short term, "things" may cost a little more until the changes are impactful, but it'll get better as we all lend our support to local business. If we pay a little more, individually, we can do so much for our local community's long-term survival, our environment, as well as our lively neighborhoods and the people we share them with.

18
Sustainable Wood

Okay, now comes the most significant change we can make as a society, in my view, to impact our environment as it relates to sustainability. The issue involves all segments of society from individual consumers, manufacturers and suppliers of raw materials, and even governments who insure any contracted services or suppliers operate with sustainability as a driving force within their companies as providing materials for all that they do. For our forests to be managed in a sustainability manner everyone has to work together if we are to succeed in this critical effort.

There are a lot of reasons to consider wood an optimal material for building and construction. Converting timber into a useable building material takes far less energy and creates minimal pollution compared to other mainstream alternatives — 5 times less energy than concrete and 6 times less than steel. Timber frame also has the lowest CO_2 cost of any commercially available building material. For every cubic meter of wood used instead of other building materials, 0.8 tons of CO_2 is saved from the atmosphere.

But to keep our industries supplied in an environmentally friendly manner requires some thought and care. How can we continue to grow and produce at a rate needed and prevent strain on forests and the environment? Lots of study has gone into sustainable woods. Most of the information below comes directly from United States Forest Service (USFS) and a primary publication by Falk, Robert H. in 2009 regarding sustainable wood use.

It's critical that we put the work in now to ensure a healthy supply for the future. This section will show you how you can do your part to make that happen. The good news is: if sustainable forestry management and harvesting practices are followed, the wood resource will be available forever.

WHAT YOU CAN DO:

Use certified sustainable wood:

First, some general statements regarding about wood's use and why it's important to use wood rather than metal or other composite materials not from sustainable sources.

Advantages of wood:

- Using wood may in some cases cost more up front, but will typically outlast other materials.
- You can get used wood furniture that may need a little work but will be well worth the effort in longevity and resale value.
- A timber frame home is a warm, comfortable and safe place in to live — plus also helps you to reduce energy costs and your carbon footprint.
- Architects and builders around the world are discovering that wood building materials are sustainable and renewable — and beautiful, too. Wood has many advantages over traditional building materials such as concrete or steel.
- Wood may be the most environmentally friendly material available for building homes or businesses.

Why using wood is so important:

- Wood is a naturally renewable resource.
- More wood is grown each year in the United States than is harvested.
- Timber is not only a recyclable resource, but it's energy-efficient to produce.
- Timber acts as a carbon store, giving it an important role to play in reducing carbon emissions. Plus, the more demand you help create by using wood from

sustainable sources, the more trees are grown that expand the benefit.
  a) Because producing timber in a sustainable way requires more long-term planning, we need as many people as possible involved and impacting the market by their purchases so suppliers can start their planning.
- If sustainable forestry management and harvesting practices are followed, the wood resource will be available forever.

Reclaimed and sustainable materials:
Using wood or buying furniture from a sustainability certified source is extremely important since, otherwise, our future generations won't have those resources to use or enjoy. "Sustainability is based on a simple principle: Everything that we need for our survival and well-being depends, either directly or indirectly, on our natural environment. To pursue sustainability is to create and maintain the conditions under which humans and nature can exist in productive harmony to support present and future generations." (Robert H. Falk, 2009)

Common varieties of sustainable wood:
Timber is usually classified as either hardwood, from broad-leafed trees such as beech and oak, or softwood from conifers like pine and fir. Fast-growing species like pine tend to be more sustainable than slow-growing trees like oak. Wood from sustainably harvested forests, sustainably harvested tree farms, and reclaimed wood are the main sources.

Reclaimed materials:
Furniture is also often made with reclaimed materials. Sustainable/industrial/recycled furniture design is an effort

to address the environmental impact of furniture products on the environment by considering all aspects of the design and manufacturing process.

Wood strength and efficiencies:

- Strength for strength, wood timber is more energy efficient.
    a) Concrete uses 5 times more energy to produce.
    b) Steel uses 6 times more energy to produce.
- Using a standard 140 mm stud timber frame system achieves U-values of 0.27 using readily available and standard insulation. Using higher performance insulation and insulation breather membranes can reduce these figures as low as 0.11.
    a) This means significant carbon savings in home day-to-day use, as well as financial benefits from lower running costs.
- Wood is not only the most widely used building material in the United States, but also one with characteristics that make it suitable for a wide range of applications.

Embodied energy:

Embodied energy refers to the quantity of energy required to harvest, mine, manufacture, and transport a material or product to the point of use. Wood, a material that requires a minimal amount of energy-based processing, has a low level of embodied energy relative to many other materials used in construction (e.g., steel, concrete, aluminum, or plastic).

- The sun provides the energy to grow trees from which wood products are produced, whereas fossil fuels are the primary energy source required in the manufacture of steel and concrete.

- Over half of the energy consumed in manufacturing wood products in the United States is from biomass (or bioenergy), which is typically produced from tree bark, sawdust, and by-products of pulping in papermaking processes.

How to find and select sustainable wood:

Forest certification programs:

Architects, product designers, material specifiers, and homeowners are increasingly asking for certified building products that are from a sustainable source. For the forest products sector, the result of this demand has been the formation of forest certification programs. While all certification programs emphasize resource sustainability, some place emphasis on issues of biodiversity, habitat protection, and indigenous peoples' rights in land management plans.

The entire Forest Certification labeling program is extremely complex, and I can't present the entire system here, but below are some examples to look for when purchasing your wood or wood products.

Global progress:

More than 50 different forest certification systems in the world today represent nearly 700 million acres of forestland and 15,000 companies involved in producing and marketing certified products. These programs represent about 8% of the global forest area and 13% of managed forests.

- From 2007 to 2008, the world's certified forest area grew by nearly 9%.
- North America has certified more than 33% of its forests.
- Europe has certified more than 50% of its forests.

- Africa and Asia, however, have certified less than 0.1%.
- Approximately 80 to 90% of the world's certified forests are located in the northern hemisphere, where two-thirds of the world's round wood is produced (UNECE 2008).

North American certification systems:

In North America, five major certification systems are used:

- Forest Stewardship Council (FSC)
- Sustainable Forestry Initiative (SFI)
- American Tree Farm System (ATFS)
- Canadian Standards Association (CSA)
- Programme for the Endorsement of Forest Certification (PEFC)

The growing trends in green building are helping drive certification in the construction market in the United States. In terms of forest acreage under certification, FSC and SFI dominate in the U.S. These two systems evolved from different perspectives of sustainability: FSC's guidelines focus on preserving natural systems while allowing for careful harvest, whereas SFI's guidelines encourage fiber productivity while allowing for conservation of resources (Howe et al. 2004).

Forest Stewardship Council (FSC):
FSC is an independent, non-governmental, not-for-profit organization established to promote responsible management of the world's forests. It's probably the most well-known forest certification program worldwide. More than 280 million acres of forest in over 79 countries worldwide are certified to FSC standards.

Types of FSC certification:
The FSC program includes two types of certification.

- Forest Management Certification applies FSC standards of responsible forestry to management of the forest land.
- Chain-of-Custody (COC) certification ensures that forest products with the FSC label can be tracked back to the certified forest from which they came.
  a) More than 14,800 COC certifications are in use by FSC members.

FSC has certified 18 certification bodies around the world. Six have offices located in the United States, including the non-profit Rainforest Alliance's SmartWood program and the for-profit Scientific Certification Systems. Both of these organizations provide up-to-date lists of FSC-certified wood suppliers in the United States.

The United States Green Building Council (USGBC) acknowledges use of FSC-certified wood and requires a minimum of 50% certified wood on a LEED (Leadership in Energy and Environmental Design) Green Building Rating System project (USGBC 2005). At this time, the USGBC does not recognize other certification systems.

How to recognize FSC-certified wood:

The labels above say it all: 100% of this product was harvested from Forest Stewardship Council-certified forests. The key word in this standard's name is stewardship, reflecting a core mission of conserving natural forest ecosystems.

It involves:

- On-the-ground auditing
- Chain-of-custody tracking
- Publicly available management plans
- Science-driven decision-making and monitoring
- Regionally sensitive treatment of ecosystems
- Protection of indigenous peoples' rights
- Fair employment practices

- Consideration of other industries that rely on a healthy forest
- Conservation of habitat
- Strict limits on pesticide use
- Prohibition of genetically modified trees
- Consultation with local experts about high-conservation-value forests
- Strict limits on conversion of forests to non-forest

Other labels identify other materials companies are allowed to use the FSC Mix label. For simplicity's sake, we'll explain the proportional method: if a mill is producing plywood with 70% FSC-certified wood content, then only 70% of the plywood coming off the mill gets the FSC Mix label. The other 30% gets no label at all. FSC requires a minimum of 70% FSC-certified forest content or post-consumer recycled content for this label to appear. The other 30% can be FSC Controlled Wood from "acceptable" sources.

Purchasing FSC Mix supports FSC-certified forestry, but since the two kinds of wood are mixed, a product that receives the FSC Mix label may not be the actual FSC-certified wood. The FSC Controlled Wood standard requires tracking to ensure any wood mixed with FSC-certified content comes from acceptable sources.

It involves:

- Legal harvesting
- Traditional and civil rights
- Protection of high-conservation-value forest
- Prohibition on plantations replacing forests
- Prohibition of genetically modified trees

Sustainable Forestry Initiative (SFI):

SFI focus:

Labels shown below represent the Sustainable Forestry Initiative (SFI). The Sustainable Forestry Initiative was established by the American Forest & Paper Association (AF&PA) in 1994 and currently certifies over 177 million acres in the United States and Canada. As of August 2009, 719 SFI COC certificates have been issued for complete chain-of-custody. This program has a strong industry focus and has been adopted by most of the major industrial forest landowners in the United States. It's based on the premise that responsible forest practices and sound business decisions can co-exist.

How to recognize SFI-certified wood:

The labels above indicate that the content was harvested from SFI-certified forests (or forests certified under similar programs recognized by SFI). The SFI Forestry Standard has many of the same elements as FSC, but its requirements are less prescriptive, companies are less transparent about their practices, and audits are arguably less rigorous. Much is left to the forester's discretion.

The SFI Forestry Standard encourages:

- On-the-ground auditing
- Chain-of-custody tracking
- Forest and soil productivity
- Some limits on converting natural forests to plantations
- Preservation of water quality
- Some protections for threatened and endangered species
- Conservation of adjacent and downstream habitats
- Some limits on pesticide use
- Prohibition of genetically modified trees
- Limitations on clear cuts that disrupt visual quality
- Measures to minimize waste

American Tree Farm System:

American Tree Farm System (ATFS), a program of the American Forest Foundation's Center for Family Forest established in 1941, is the oldest forest certification program and focuses on private family forest landowners in the United States. Currently, ATFS has certified 24 million acres of privately owned forestland of 90,473 family forest owners in 46 states.

ATFS has established standards and guidelines property owners must meet to become a Certified Tree Farm. Under these standards and guidelines, private forest owners must develop a management plan based on strict environmental standards and pass an inspection by an ATFS volunteer forester every five years.

Canadian Standards Association:

The Canadian Standards Association (CSA), a non-profit

organization, has developed over 2,000 standards for a variety of industries. CSA first published Canada's National Standard for Sustainable Forest Management (SFM) CAN/CSA-Z809 in 1996. The SFM program has four components: the SFM Standard itself, a COC pro- gram, product marking, and the CSA International Forest Products Group, which promotes the program. The CSA Standard has been adopted by the major industrial forestland managers in Canada. AF&PA has also accepted the CSA Standard as the "functional equivalent of the SFI Standard" (Fernholz et al. 2005). As of June 2007, approximately 60% (198 million acres) of Canadian forests were certified under the CAN/CSA-Z809 SFM Standard.

WHY IT MATTERS:

Positive environmental impacts:

- Wood is effectively a carbon-neutral material (even allowing for transport).
- Timber frame has the lowest CO_2 cost of any commercially available building material.
- For every cubic meter of wood used instead of other building materials, 0.8 tons of CO_2 is saved from the atmosphere.
- Processing timber is also not a gas-guzzling procedure like other manufacturing processes.
    a) 77% of the energy used in the production of wood products comes from wood residues and recovered wood.
    b) Converting timber into a useable building material takes far less energy and creates minimal pollution compared to other mainstream alternatives such as aluminum, steel, concrete and brick.
- Waste and "end of life" wood can be easily recycled.

- Wood is a green building material
  a) Over the past decade, the concept of green building has entered the mainstream and the public has become increasingly aware of the potential environmental benefits of this alternative to conventional construction.
  b) Much of the focus of green building is on reducing the energy consumption of a building (such as better insulation and more efficient appliances and HVAC systems) and reducing negative human health impacts (such as controlled ventilation and humidity to reduce mold growth).
- Carbon contained in wood products currently in use and as wood debris in landfills is estimated at 2.7 billion tons and accumulates at a rate of 60 million tons per year.
- In the United States, slightly more than half of the wood harvested in the forest is used in construction.

Falk concludes that it's clear the green building movement is here to stay and will undoubtedly grow in the future. This can be good for the wood industry, because there's a positive and convincing story to tell about wood as a sustainable and environmentally preferable material. By providing the green building community with science-based facts about sustainability, embodied-energy, and carbon impact, wood can stand out as the greenest of building materials.

I know the information above is a lot to absorb if you've never been aware of the sustainability certification system and its objectives and processes. But, please know that if you take the effort to utilize wood from sustainable sources, make your local businesses and governments aware of your concern you and your family can really help the environment, conserve our worldwide resources, and provide our future generations with a building material source that will truly impact the survivability of this planet.

CONCLUSIONS
What If We Don't Make Changes?

"Human beings can always be relied upon to exert, with vigor, their God-given right to be stupid."
~ DEAN KOONTZ

Simply, the planet will die.

When it will die or reach the point where habitation is more than a little tough and is a contested debate. Information that follows provides some specifics that should help you explain to others why our efforts are so critical. Whether it's 50 or 500 years shouldn't affect your motivation; it's clear from all the data that it will happen if we don't take steps to change what we're doing individually, as communities, industry and governments. The plain truth is: we've reached a situation of "Clear and Present Danger" because we're killing mother earth and depleting her natural resources while making it continuously harder and harder for her to rebound.

The earth is currently facing a lot of environmental concerns. Sadly, humans have helped cause many of these environmental problems through population growth, wasteful and unsustainable resource use, systemic poverty, failure to include harmful environmental costs of goods and services in market prices, and insufficient knowledge of how nature works. Dangers like global warming, depletion of natural resources, decreased water quality, acid rain, air pollution, waste disposal, ozone layer depletion, water pollution, and climate change affect every human, animal and nation on this planet.

Fortunately, we can help to delay and mitigate the impacts with some simple changes in habits and lifestyle, including those you've read in this book.

We can also choose to do nothing, but not when and how we reap the consequences.

These are just a few of the potential results of apathy.

DEPLETED RESOURCES:

Impacts are many, but a few of the numbers and impacts are as follows:

- Every year, we extract an estimated 55 billion tons of fossil energy, minerals, metals and bio mass from the earth.
- The world has already lost 80% of its forests, and we're continually losing them at a rate of 375 km2 per day.
- At the current rate of deforestation, 5-10% of tropical forest species will become extinct every decade.
- Every hour, 1,692 acres of productive dry land become desert.
- 27% of our coral reefs have been destroyed. If the rate continues, remaining 60% will be gone in 30 years.
- We're using up 50% more natural resources than the earth can provide. At our current population, we need 1.5 Earths, which we don't have.
- The United States makes up less than 5% of the population on earth, yet we easily consume over 30% of its resources. While we humans would appear to be doing well, spreading our population [117] like wildfire across the globe, the diminishing resources and other life forms on the planet tell a different story. "We are in the midst of a mass extinction, an event not seen since the disappearance of the dinosaurs, 65 million years ago," says the Worldwatch Institute.[118]

POLLUTION AND LANDFILLS:

- Americans dump 16 tons of sewage into their waters every minute.

- Billions of plastic bags (119) are made each year. Of these bags, one hundred billion are thrown away, according to Worldwatch Institute, with less than one percent finding their way into a recycle bin. The end result is approximately one billion birds and mammals dying each year by the ingestion of plastic.
- Landfills generated by the large amount of waste that's generated by households, industries and healthcare centers every day produce toxic gases that threaten our environment and could prove fatal for humans and animals.

GARBAGE ISLANDS:

Developed nations are infamous for creating an unreasonable amount of waste or junk and dumping it in the oceans and seas, as well as less developed nations or nations that have decided to make money by accepting other waste.

- We have a garbage island floating in our ocean — mostly comprised of plastics — the size of India, Europe and Mexico combined.
- There are several garbage islands in our oceans, and the most "popular" of them is the Great Pacific Garbage Patch. Its size cannot be pinned down since it's constantly growing.
- The garbage island is made up of pelagic plastic, debris and chemical sludge that got trapped in the North Pacific Gyre. It's a collection of waste coming from the surrounding countries, brought to this location by the currents.
- The Great Pacific Garbage Patch is composed of 2 landfill masses — the Eastern and Western Garbage Patch.
- The Eastern Garbage patch can be found between Hawaii and California and scientists have estimated its size to be twice that of Texas.

- The Western Garbage Patch can be found east of Japan and west of Hawaii. These 2 patches, connected by the Subtropical Convergence Zone, are the biggest plastic landfills in the ocean today.
- The Atlantic Ocean also has a garbage patch, in the Sargasso Sea.
- Aside from the Pacific and Atlantic Oceans, plastic landfills are turning up in other major tropical oceanic gyres in the world.

DESTRUCTION OF CORAL REEFS:

- Pollution: Corals need clean water to live. The garbage that we throw in our oceans affects the quality of the water, not to mention the fertilizers, pesticides, sewage, oil pollution and other man made toxins dumped directly or carried into our oceans by rain or rivers that are poisoning the reefs.
- Climate Change: Global warming has caused an increase in coral bleaching, and it can only get worse. 80% of the added heat in our climate is being absorbed by the ocean, and corals cannot survive in waters with too high temperatures.
- Tourism: Standing on corals and touching them, as well as boat propellers, snorkeling, dropping anchors on the reef can cause damage that will take millennia to mend, if ever.
- Destructive Fishing Methods: Cyanide fishing kill polyps and algae. It requires the fishermen to break the corals in order to get the fish that they stunned. Muro ami is the act of banging the reef with sticks. These practices have been banned, including blast and dynamite fishing.
- Commercial Traps: Lobster traps are light enough to be moved by the current but heavy enough to damage a coral when it comes in contact with one.

- Too Much Sediment: The sediment, which comes from farms or construction sites near the ocean, kills corals by blocking the sunlight or clogging their mouths. (Yes, they have mouths)

Okay, that's enough of the scary reasons we need to do something. The problems are real the facts are there. A reference list is provided at the end of this book for more information. I encourage you to do more investigation on your own and keep trying to make a better world.

Why You Should Act:
There are countless reasons why you should get involved and take action to develop a more environmentally friendly and sustainable lifestyle. Most of them will differ from person to person, based on a variety of factors. But, in the end, we all share the most critical one: our future and that of our kids and grandkids depends on us doing the right thing.

More damning evidence of our harm to the earth and its environment and ecosystems keeps emerging, and the worsening of our environment is more and more visible. We can't turn our heads and ignore the problem any longer. With a clear and present danger to our very survival, our only hope is to act, do it quickly, and teach the next generation a better way.

We have a chance to give the generations to come the best gift of all, the hope of many more tomorrows and a healthy earth to sustain them.

Our future will only come when we can motivate, educate and involve our young people in this effort. So, even if you can only follow through on a few suggestions you've read here, please involve the kids in your life and explain the what's, whys, and how's as much as possible. They'll not only enjoy your company, but be inspired by your belief in protecting and improving this biosphere we live in and the earth we dwell on.

Thank you for all you do to make a difference.

☐ REFERENCES

1. Biology Online, 2018, available from
   https://www.biologyonline.com/dictionary/biosphere

2. The World Counts, 2018, available from
   http://www.theworldcounts.com/stories/environmental-degradation-facts

3. Conserve Energy Future, 2018, available from
   https://www.conserve-energy-future.com/top-25-environmental-concerns.php

4. Treehugger, 2018, available from
   https://www.treehugger.com/clean-technology/20-gut-wrenching-statistics-about-the-destruction-of-the-planet-and-those-living-upon-it.html

5. United States Environmental Protection Agency, 2018, avail- able from https://www.epa.gov/sustainability/learn-about-sustainability#what

6. Worldwatch Institute, 2018, available from
   http://www.worldwatch.org/

7. Advanced Placement Environmental Science reference materials.

8. Care2, 2018, available from
   https://www.care2.com/greenliving/5-reasons-why-people-dont-recycle-and-5-reasons-they-should.html

9. The Guardian (2002) available from
   https://www.theguardian.com/environment/2002/aug/22/worldsummit2002.earth21

10. Natures Path, Things You Can Do To Save The Environment, available from https://www.naturespath.com/en-us/blog/nine-things-you-can-do-to-save-the-environment/

11. How Stuff Works, Katie Lambert, 10 Things You Can Do to Help Save the Earth, available from https://science.howstuffworks.com/environmental/green-science/save-earth-top-ten.htm

12. Bag It The Movie, Information available from https://tubitv.com/movies/317275/bag_it

13. Department of Natural Resources, 100 Ways You Can Save the Earth, available from http://infohouse.p2ric.org/ref/15/14300.pdf

14. Science Based Life, 100 Ways To Personally Help the Environment, available from https://sciencebasedlife.wordpress.com/2010/12/30/10-ways-to-personally-help-the-environment

15. One Cent At A Time, 101 Ways You Can Save Energy and Save Environment, available from http://onecentatatime.com/101-ways-to-save-energy-environment/

16. Mother Earth News, The Plastic Bag Problems, available from https://www.motherearthnews.com/nature-and-environment/environmental-policy/plastic-bag-problem-ze0z1302zwar

17. Sea Turtle Conservancy, Information About Sea Turtles: Threats from Marine Debris, available from https://conserveturtles.org/information-sea-turtles-threats-marine-debris/

18. Plastic Pollution Coalition, Why Is Plastic Harmful, available from https://plasticpollutioncoalition.zendesk.com/hc/en-us/articles/222813127-Why-is-plastic-harmful-

19. Arizona State University, The Biodesign Institute, Perils of Plastics: Risk To Human Health and the Environment, available from https://biodesign.asu.edu/news/perils-plastics-risks-human-health-and-environment

20. The New York Times, 2018, Joseph Curtin (Opinion Contributor), Let's Bag Plastic Bags, available from https://www.nytimes.com/2018/03/03/opinion/sunday/plastic-bags-pollution-oceans.html

21. New York State, 2018, An Analysis of the Impact of Single-Use Plastic Bags, available from https://www.dec.ny.gov/docs/materials_minerals_pdf/dplasticbagreport2017.pdf

22. Center for Biological Diversity, 2018, available from https://www.biologicaldiversity.org/programs/population_and_sustainability/sustainability/plastic_bag_facts.html

23. Virginia Department of Environmental Quality, Recycling Aluminum Cans Makes Cents, available from https://www.deq.virginia.gov/Portals/0/DEQ/ConnectWithDEQ/EnvironmentalInformation/VirginiaNaturally/LAT2009/LAT2009_RecyclingCans.pdf

24. US Environmental Protection Agency, Municipal Solid Waste Generation, Recycling and Disposal in the United States: Facts and Figures for 2012, available from https://www.epa.gov/sites/production/files/2015-09/documents/2012_msw_fs.pdf

25. City of Miami Florida, Interesting Facts About Recycling, available from http://www.miamiokla.net/DocumentCenter/View/379/RECYCLING-FACTS?bidId=

26. Waste-free Mail, 2018, Saving the Planet One Mailing At A Time, FAQ, available from http://www.wastefreemail.com/faq.html

27. University of Southern Indiana, Paper Recycling Facts, available from https://www.usi.edu/recycle/paper-recycling-facts/

28. Plastics Recycling Update 2018, Jerry Powell, 2018 Recycling Market Update, available from https://resource-recycling.com/recycling/author/jerry-powell/

29. US Department of Energy, Thermostats, available from https://www.energy.gov/energysaver/thermostats

30. Direct Energy, 2018, Josh Crank, How Much Can You Save By Adjusting Your Thermostat, available from http://www.directenergy.com/blog/how-much-can-you-save-by-adjusting-your-thermostat/

31. One Green Planet, What's The Problem With Plastic Bottles, available from https://www.onegreenplanet.org/animalsandnature/whats-the-problem-with-plastic-bottles/

32. US Department of Energy, 2009, 15 Ways To Save On Your Water Bill, available from https://www.energy.gov/energysaver/articles/15-ways-save-your-water-heating-bill

33. Rotoplas, 2017, 10 Uses For Collected Rainwater, available from http://rotoplasusa.com/uses-collected-rainwater/

34. Texas Water Development Board, The Texas Manual on Rainwater Harvesting, Third Edition, available from http://www.twdb.texas.gov/publications/brochures/conservati on/doc/RainwaterHarvestingManual_3rdedition.pdf

35. Green Mom, Why Styrofoam Is So Bad For The Environment, available from https://green-mom.com/styrofoam-bad-environment/

36. Call 811, Do I Really Need To Call 811, available from http://call811.com/before-you-dig/do-i-really-need-call

37. Rutgers University, Adrienne Miller, Sheila Mohazzebi, Samantha Pasewark with Julie M. Fagan, Ph.D., Styrofoam: More Harmful than Helpful, available from https://rucore.libraries.rutgers.edu/rutgers-lib/38329/PDF/1/play/

38. Wikihow, How to Field Strip a Cigarette, available from https://www.wikihow.com/Field-Strip-a-Cigarette

39. US National Library of Medicine, National Institutes of Health, 2009, Cigarettes Butts and the Case for an Environmental Policy on Hazardous Cigarette Waste, available from https://www.ncbi.nlm.nih.gov/pmc/articles/PMC2697937

40. World Health Organization, Tobacco and Its Environmental Impact: an overview, available from http://www.who.int/tobacco/publications/environmental-impact-overview/en/

41. RSPCA, 2017, What is the most humane way to kill pest rats and mice, available from http://kb.rspca.org.au/what-is-the-most-humane-way-to-kill-pest-rats-and-mice_139.html

42. Dengarden, 2017, Michael Kismet, 5 Simple Ways to Get Rid of Mice Without Killing Them, available from https://dengarden.com/pest-control/5-Simple-Ways-to-get-rid-of-Mice-without-Killing-Them

43. Department of the City and County of San Francisco, 2012, Chris A. Geiger, Caroline Cox, Pest Prevention by Design — Authoritative Guidelines for Designing Pests Out of Structures, available from https://sfenvironment.org/

44. How to Get Rid of Mice, Top 3 Humane Mouse Traps Reviewed — Pros and Cons of the Best, available from http://how-to-get-rid-of-mice.com/humane-mouse-traps/

45. How to Get Rid of Mice, Natural Home Remedies To Get Rid Of Mice, available from https://how-to-get-rid-of-mice.com/natural-home-remedies/

46. How to Get Rid of Mice, 10 Facts about Mice to Help You Get Rid of Them, available from https://how-to-get-rid-of-mice.com/facts-about-mice/

47. How to Get Rid of Mice, How Much Does a Mouse Exterminator Cost, available from https://how-to-get-rid-of-mice.com/mouse-exterminator-cost/

48. US Environmental Protection Agency, 2015, Advancing Sustainable Materials Management: 2015 Fact Sheet, available from https://www.epa.gov/facts-and-figures-about-materials-waste-and-recycling/advancing-sustainable-materials-management-0

49. Carroll County, Maryland, 2018, A guide to Waste Management & Recycling In Carroll County Maryland, available from http://ccgovernment.carr.org/ccg/solidwaste/default.asp

50. Center for Disease Control (CDC), Field Identification OF Domestic Rodents, available from https://www.cdc.gov

51. Center for Disease Control (CDC), Ecology and Transmission, available from https://www.cdc.gov/plague/transmission/index.htm

52. American Veterinary Medical Association (AVMA), 2013, AVMA Guidelines for the Euthanasia of Animals: 2013 Edition, available from https://www.avma.org/KB/Policies/Documents/euthanasia.pdf

53. Lawn Care Academy, Complete Tree Planting Instructions, The No Regrets Tree Guide, available from https://www.lawn-care-academy.com/tree-planting-instructions.html

54. Ocean Conservancy, Trash Free Seas, available from https://oceanconservancy.org/trash-free-seas/

55. United States Department of Agriculture (USDA), Forest Service, Inventories Of Woody Residues And Solid Woos Waste In The United States, David B. McKeever, available from www.fs.fed.us

56. US Environmental Protection Agency, Managing, Reusing, and Recycling Used Oil, available from https://www.epa.gov/recycle/managing-reusing-and-recycling-used-oil

57. Child Likes, Remote Power Transfer — the end of batteries, available from http://www.childlikes.com/battery.htm

58. Raw Materials Company Inc., How to Prepare Batteries for Recycling available from https://www.rawmaterials.com/page/education/prepare-batteries/

59. Gemeinhardt, Ronald, L, 2016, Texas Used Oil Management: A Practical Guide for Environmental Compliance, Graduate Thesis, Green Mountain College

60. Energy Sage, 2018 Most Energy Efficient Appliance, available from https://www.energysage.com/energy-efficiency/costs-benefits/energy-star-rebates/

61. Department of Energy, When to Turn Off Your Lights, available from https://www.energy.gov/energysaver/when-turn-your-lights

62. Texas Government Land Office, Adopt A Beach, available from http://www.glo.texas.gov/adopt-a-beach/index.html

63. Inc. Com, Bill Murphy Jr., 7 Steps to Persuade Your Boss to Let You Work From Home, available from https://www.inc.com/bill-murphy-jr/7-steps-to-convince-your-boss-to-let-you-work-from-home.html

64. Earth911, Recycler Centers Look-up, available from https://search.earth911.com/

65. Minnesota Pollution Control Agency, Changing your oil: An earth-friendly guide for do-it-yourselfers, available from https://www.pca.state.mn.us/living-green/changing-your-oil-earth-friendly-guide-do-it-yourselfers

66. Illinois Environmental Protection Agency, 2005, How Do I Manage My Used Oil and Used Oil Filters? Available from http://www.epa.illinois.gov/

67. American Public Transportation Association, Calculate Your Savings by Riding Public Transportation, available from https://www.apta.com/resources/aboutpt/pages/transitcalculator.aspx

68. Public Transit in Your Community, Fuel Savings Calculator, available from https://www.publictransportation.org/tools/fuelsavings/Pages/default.aspx

69. Blue Ocean Society for Marine Conservation, available from https://www.blueoceansociety.org

70. USDA, US Forest Service, Forest Products Journal Vol. 59, No. 9, available from https://www.fpl.fs.fed.us/documnts/pdf2009/fpl_2009_falk001.pdf

71. American Forest Foundation, Wood: A Good Choice for Energy Efficiency and the Environment, available from https://www.forestfoundation.org/wood—a-good-choice-for-energy-efficiency-and-the-environment

72. US Environmental Protection Agency, Sustainable Materials Management (SMM) Electronics Reuse and Recycling Forum, available from https://www.epa.gov/smm-electronics

73. Building Green, What These Forestry Labels Really Mean, available from https://www.buildinggreen.com/infographic/what-these-forestry-labels-really-mean

74. United Nations University, 2017, The Global E-waste Monitor 2017, available from https://collections.unu.edu/eserv/UNU:6341/Global-E-waste_Monitor_2017__electronic_single_pages_.pdf

75. Georgia Environmental Compliance Assistance Program, Electronic Waste, available from http://www.gecap.org Michigan State University, 2010, Why Buy Local—An Assessment of the Economic Advantages of Shopping at Locally Owned Businesses, Principal Author Nandi Robinson, available from http://www.ced.msu.edu/upload/reports/why%20buy%20local.pdf

76. Century Link Brand Voice, 2017, 5 Benefits Of Shopping Locally On Small Business Saturday, available from https://www.forbes.com/sites/centurylink/2017/11/20/5-benefits-of-shopping-locally-on-small-business-saturday

77. US Environmental Protection Agency, Safer Choice Program, available from www.epa.gov/saferchoice

78. Bob Villa, Joe Provey, What to Do With Old Paint, available from https://www.bobvila.com/articles/what-to-do-with-old-paint/

79. Grand Traverse County, Household Paint Disposal Guide, available from www.recyclesmart.info

80. Healthline, Natural Mosquito Repellents, available from https://www.healthline.com/health/kinds-of-natural-mosquito-repellant

81. PestWiki, 2018, 6 Effective Home Remedies to Kill Roaches (Naturally), available from https://www.pestwiki.com/best-roach-home-remedies/

82. American Mosquito Control Association, Frequently Asked Questions, available from https://www.mosquito.org/page/faq

83. US Environmental Protection Agency, 2017, Citizen's Guide to Pest Control and Pesticide Safety, available from https://www.epa.gov/safepestcontrol/citizens-guide-pest-control-and-pesticide-safety

84. Center for Disease Control (CDC), Healthy Housing Reference Manual Chapter 4: Disease Vectors and Pests, available from https://www.cdc.gov/nceh/publications/books/housing/cha04.htm

85. National Oceanic And Atmospheric Administration, Beach And Waterway Cleanups, available from www.marinedebris.noaa.gov

86. Well Organization, 10 Advantages of Buying Local, available from https://well.org/conscious-consumers/10-advantages-of-buying-local

☐

Notes / Web Site References

1. https://www.energy.gov/energysaver/articles/program-your-thermostat- fall- and-winter-savings
2. https://www.energy.gov/energysaver/thermostats
3. http://www.energystar.gov/index.cfm?c=clotheswash.pr_clotheswashers
4. http://www.wmnorthwest.com/guidelines/plasticvspaper.htm
5. https://www.dec.ny.gov/chemical/112291.html
6. https://www.unenvironment.org/news-and-stories/press-release/un- declares-war- ocean-plastic
7. http://www3.weforum.org/docs/WEF_The_New_Plastics_Economy.pdf
8. http://ec.europa.eu/environment/circular-economy/pdf/plastics- strategy.pdf
9. http://journals.plos.org/plosone/article?id=10.1371/journal. pone.0111913
10. http://www.iflscience.com/health-and-medicine/seafood-eaters-may-be-ingesting-up- to-11000-microplastic-particles-a-year/
11. http://time.com/4928759/plastic-fiber-tap-water-study/
12. https://www.usatoday.com/story/news/health/2017/09/06/94-u-s- tap-water-contaminated-plastic-fibers-including-faucets-trump-power/636662001/
13. http://www.sprep.org/attachments/Publications/FactSheet/ plasticbags.pdf
14. http://www.wmnorthwest.com/guidelines/plasticvspaper.htm
15. http://www.sprep.org/attachments/Publications/FactSheet/plasticbags. pdf
16. http://www.environmentmassachusetts.org/sites/environment /files/reports/BagBanFactSheet _0.pdf
17. http://www.nrdc.org/media/2008/080109
18. http://www.wmnorthwest.com/guidelines/plasticvspaper.htm
19. http://www.sprep.org/attachments/Publications/FactSheet/plasticbags.pdf
20. http://www.worldwatch.org/node/5565
21. http://www.wmnorthwest.com/guidelines/plasticvspaper.htm
22. http://ecowatch.com/2013/08/06/the-danger-of-plastic-bags-to-marine-life
23. http://www.treehugger.com/clean-water/the-us-consumes-1500-plastic-water-bottles-every-second-a-fact-by-watershed.html
24. https://www.worldatlas.com/articles/top-bottled-water-consuming- countries.html
25. http://en.wikipedia.org/wiki/Bisphenol_A
26. http://en.wikipedia.org/wiki/Bisphenol_A

27. http://www.scientificamerican.com/article.cfm?id=just-how-harmful-are-bisphenol-a- plastics

28. http://www.environmentalhealthnews.org/ehs/newscience/bpa-crosses-placenta-is-active-form-in-fetus/

29. http://www.bagitmovie.com/

30. 30 https://www.nbcnewyork.com/news/local/Several-Brands-Bottled-Water-Recalled-Niagara-Bottling-308966101.html?fbclid=IwAR297T_gBaJnHfgDB6BzSYLRd52wAWNgcxb9fT1gupU_CuQ2tLKHnWeWXkc

31. https://www.plasticsmakeitpossible.com/plastics-recycling/how-to-recycle/at-home/keep-your-top-on/

32. http://discovermagazine.com/2009/jul-aug/06-when-recycling-is-bad-for-the-environment

33. http://www.epa.gov/

34. http://www.inchem.org/

35. http://www.inchem.org/documents/iarc/vol82/82-07.html

36. http://www.groundswell.org/map-which-cities-have-banned-plastic-foam/

37. http://www.greenecoservices.com/deadly-litter-and-car-accidents/

38. http://www.therightfuton.com/Modern-Convertible-Sofa-Sleeper.html

39. http://www.therightfuton.com/

40. http://www.therightfuton.com/Cheap-Futon-Mattress-Cover.html

41. http://www.epa.gov/osw/nonhaz/municipal/pubs/msw2009rpt.pdf

42. http://earth911.com/recycling/household/home-and-office-furniture/facts-about-household-furniture/

43. https://www.nfpa.org/News-and-Research/Fire-statistics-and-reports/Fire-statistics/Fire-causes/Smoking-Materials

44. https://www.verywellmind.com/harmful-chemicals-in-cigarettes-and-cigarette-smoke-2824715

45. http://www.wpro.who.int/china/mediacentre/factsheets/tobacco/en/

46. https://www.verywellmind.com/harmful-chemicals-in-cigarettes-and-cigarette-smoke-2824715

47. https://www.verywellmind.com/tar-in-cigarettes-2824718

48. https://www.verywellmind.com/cigarette-additives-2824737

49. http://www.longwood.edu/cleanva/

50. https://www.nfpa.org/News-and-Research/Fire-statistics-and-rep

51. https://www.ncbi.nlm.nih.gov/pmc/articles/PMC4307979/orts/Fire- st

52. https://www.energy.gov/energysaver/save-electricity-and-fuel/lighting-choices-save-you-money/incandescent-lighting

53. https://www.energy.gov/energysaver/save-electricity-and-fuel/lighting-choices-save-you-money/incandescent-lighting

54. https://www.energy.gov/energysaver/save-electricity-and-fuel/lighting-choices-save-you-money/fluorescent-lightingatistics/Fire-causes/Smoking-Materials

55. https://www.energy.gov/energysaver/save-electricity-and-fuel/lighting-choices-save-you-money/replacing-lightbulbs-and

56. https://whatis.techtarget.com/definition/semiconductor

57. https://whatis.techtarget.com/definition/current

58. https://www.energy.gov/energysaver/save-electricity-and-fuel/lighting-choices-save-you-money/lighting-controls

59. https://www.energy.gov/energysaver/save-electricity-and-fuel/lighting-choices-save-you-money/lighting-controls

60. https://www.myledlightingguide.com/blog-save-money-on-your-utility-bill

61. https://www.energystar.gov/products/appliances

62. http://news.energysage.com/residential-electricity-prices-going-up-or-d

63. http://planetgreen.discovery.com/tech-transport/carbon-emissions-increase.htmlown/

64. http://www.carmel.in.gov/Home/ShowDocument?id=179

65. http://www.blueoceansociety.org/wp-content/uploads/2016/03/BOS-Beach-Cleanup-Guidelines.pdf

66. http://www.ibtimes.co.uk/there-are-more-gadgets-there-are-people-world-1468947

67. http://www.triplepundit.com/special/circular-economy-and-green-electronics/our-connected-mobile-recycled-and-green-future/

68. http://e-stewards.org/find-a-recycler/

69. http://www.epa.gov/recycle/electronics-donation-and-recycling

70. http://www.triplepundit.com/special/circular-economy-and-green-electronics/our-connected-mobile-recycled-and-green-future/

71. https://www.ecycleclearinghouse.org/contentpage.aspx?pageid=10

72. http://www.electronicsrecycling.org/?page_id=37

73. http://www.ibtimes.co.uk/there-are-more-gadgets-there-are-people-w

74. http://content.time.com/time/photogallery/0,29307,1870162,00.html

75. http://www.epa.gov/recycle/electronics-donation-and-recyclingorld-1468947

76. http://www.businessinsider.com/the-lesser-known-facts-about-e-waste-re

77. https://academic.oup.com/cjres/article-abstract/5/1/149/326109?redirectedFrom=fulltextcycling-2012-10

78. https://www.healthline.com/health/yellow-fever

79. https://www.healthline.com/health/food-nutrition/almond-oil

80. http://www.fda.gov/Cosmetics/ProductsIngredients/Products/ucm127054.htm

81. http://mosquito.ifas.ufl.edu/Mosquito_Repellents.htm

82. http://amzn.to/2tWYK3x

83. https://wwwnc.cdc.gov/travel/yellowbook/2016/the-pre-travel-consultation/protection-against-mosquitoes-ticks-other-arthropods

84. https://www.healthline.com/health/9-ways-eucalyptus-oil-can-help

85. http://www.ncbi.nlm.nih.gov/pubmed/24772681

86. http://amzn.to/2uUU8jc

87. https://www.ncbi.nlm.nih.gov/pubmed/12542193

88. https://www.healthline.com/health/what-lavender-can-do-for-you

89. http://pubs.acs.org/doi/abs/10.1021/jf0497152

90. http://amzn.to/2tWOdp8

91. https://www.healthline.com/health/health-benefits-of-thyme

92. http://www.ncbi.nlm.nih.gov/pubmed/12542193

93. http://amzn.to/2vGTNOA

94. http://malariajournal.biomedcentral.com/articles/10.1186/1475-2875-10-S1-S11

95. http://www.ncbi.nlm.nih.gov/pubmed/24449446

96. http://www.sciencedaily.com/releases/2001/08/010828075659.htm

97. http://mosquito.ifas.ufl.edu/Mosquito_Repellents.htm

98. https://www.amazon.com/Blocker-Organic-Insect-Repellent-Spray/dp/B003CTUXEO

99. http://amzn.to/2uuSbZB

100. http://www.ncbi.nlm.nih.gov/pubmed/22299433

101. http://amzn.to/2eLwbUM

102. http://malariajournal.biomedcentral.com/articles/10.1186/1475-2875-10-S1-S11

103. https://www.healthline.com/health/tea-tree-oil

104. http://amzn.to/2tX6qmv

105. http://onlinelibrary.wiley.com/doi/10.1111/j.1440-6055.2009.00736.x/full

106. https://www.healthline.com/health/fly-bites

107. http://amzn.to/2vGnMWK

108. http://www.pestwiki.com/cockroach-facts/

109. http://www.thehealthsite.com/topics/skin-rash/

110. http://www.thehealthsite.com/diseases-conditions/4-tips-to-prevent-asthma-attacks-during-the-winter/

111. http://www.about-salmonella.com/

112. http://www.pestwiki.com/boric-acid-for-roaches/

113. https://www.thespruce.com/where-should-i-put-rat-traps-2656748

114. https://www.thespruce.com/rats-and-mice-traps-and-baits-2656477

115. https://www.thespruce.com/the-difference-between-rats-and-mice-2656563

116. http://www.cdc.gov/rodents/

117. https://www.treehugger.com/files/2009/03/better-health-reduce-population-poverty.php?daylife=1&dcitc=daylife-article

118. http://www.worldwatch.org/

119. https://www.mnn.com/lifestyle/recycling

ABOUT THE AUTHOR

MR. RONALD L. GEMEINHARDT, MSES, (Masters of Science in Environmental Studies) has 45 years of experience in the petroleum industry beginning with USAF POL Specialists for four years. With a significant Environmental Health and Safety (EH&S) compliance support background in Refining, Distribution, Pipeline Operations, and Retail. With a focused in-depth background in waste management including recycling and used oil management. Environmental support areas included Incident Command System (ICS) application, within the Shell Oil Company Response Group and three years on the Deepwater Horizon Incident. As a Shell Oil Team Leader responsibilities included support to Distribution, Pipeline and Retail residual management with a significant focus on Resource Conservation Recovery Act (RCRA) compliance. Primary activities include, staff management, waste management, incident management and investigation support, remediation, litigation support, plan writing and review, training development and implementation, and on-site training. Facility operational support and on-site management background in Safety, Health, Environmental Compliance related to Air, Water, Waste, Comprehensive Environmental Response, Compensation, and Liability Act (CERCLA), Toxic Substances Control Act (TSCA), site remediation, and Department of Transportation (DOT), Fuels, Used Oil collection and processing, and Finished Lubricants production compliance management. Refining and Lubricant Compounding QA/QC laboratory management. Specialized Trainer Highlights include Terminal / Pipeline / Retail Environmental Compliance Training.

www.ingramcontent.com/pod-product-compliance
Lightning Source LLC
Chambersburg PA
CBHW071405210526
45465CB00001B/262